New Algorithms, Architectures and Applications for Reconfigurable Computing

New Algorithms, Architectures and Applications for Reconfigurable Computing

Edited by

Patrick Lysaght
Xilinx, San Jose, USA

and

Wolfgang Rosenstiel
University of Tübingen, Germany

 Springer

A C.I.P. Catalogue record for this book is available from the Library of Congress.

ISBN-13 978-1-4419-5264-6 (PB)
ISBN-13 978-1-4020-3128-1 (e-book)

Published by Springer,
P.O. Box 17, 3300 AA Dordrecht, The Netherlands.

Sold and distributed in North, Central and South America
by Springer,
101 Philip Drive, Norwell, MA 02061, U.S.A.

In all other countries, sold and distributed
by Springer,
P.O. Box 322, 3300 AH Dordrecht, The Netherlands.

Printed on acid-free paper

Printed in the Netherlands.

Contents

Methodologies and Tools

Applications

viii

Introduction

In describing the progression of the semiconductor industry over the last several decades, Dr. Tsugio Makimoto proposed a model that has subsequently come to be known as Makimoto's Wave. The model consists of cycles that alternate between product standardization and customization. According to the model, we are now firmly in the middle of a field programmable cycle that is characterized by *standardization in manufacturing and customization in application*. Field programmable gate arrays (FPGAs) are the standard bearers of this new age. At a time when only the largest companies can afford to fully exploit state-of-the-art silicon, FPGAs bring the benefits of deep submicron and nano-scale integration to even the smallest teams. It is fitting then that we are witnessing a huge surge in the amount of research being conducted worldwide into field programmable logic and its related technologies. This book was inspired by the desire to compile a selection of the best of the current research efforts and present it in an accessible way to a wide audience.

The book consists of a collection of invited contributions from the authors of some of the best papers from the Design, Automation, and Test conference in Europe (DATE'03) and the Field Programmable Logic conference (FPL'03). The contributions are organized into twenty-four chapters. In all, eighty-five authors have described their latest research, making this volume a unique record of the state-of-the-art in FPGA-related research.

The twenty-four chapters are organized into three main categories. The first five chapters address hard and soft architectures. Chapters 6 to 13 are concerned with design methodology and tools. The final eleven chapters are devoted to applications, of which chapters 14 to 21 focus on networking, encryption and security. Some of the research reported here defies simple classification so there is more than one example of work that could appear in more than one category. Nonetheless, the classification scheme does impose a useful order on the work.

Architectures

In the opening chapter, Schmit presents an insightful analysis of the challenges facing architects of island-style FPGAs in scaling interconnect to match increasing netlist complexity. His approach is based on the introduction of additional wiring in the *logical* third and fourth dimensions and then mapping these additional routes into the two-dimensional silicon. The benefits the author claims for this strategy include longer tile life, better evolution of CAD tools and a unification of the analysis of time-multiplexed FPGAs.

The next two chapters introduce two new coarse-grained architectures. The motivation for coarse-grained architectures includes better computational and power efficiency, and improved opportunities for the automatic compilation of designs from higher-level design representations. Mei and colleagues propose ADRES (Architecture for Dynamically Reconfigurable Embedded System) which is a novel integration of a VLIW processor and a coarse-grained reconfigurable matrix that combine to form a single architecture with two virtual, functional views. They also present a novel modulo scheduling technique which they claim can map kernel loops on to the matrix efficiently by simultaneously solving the placement, routing and scheduling problems.

Chapter 3 describes the structures and principles of the stream-based, reconfigurable PACT/XPP architectures. The status of an adaptive System-on-Chip (SoC) integration is given, consisting of a SPARC-compatible LEON processor-core, a coarse-grain XPParray, and an Amba-based communication interfaces.

In chapters 4 and 5 two soft architectures are proposed. Bidarte et al describe their *system on reconfigurable chip* (SORC). This is a core based system-on-chip architecture that they have verified with two industrial control type applications. They considered a number of interconnection networks before selecting the Wishbone interconnection architecture for portable IP cores that is available from OpenCores.org. The second core-based architecture, proposed by Roma et al, is specialized for motion estimation applications. The authors describe a customizable core-based architecture for real-time motion estimation in FPGAs using a two-dimensional array processor with extensive parallelism and pipelining. By applying sub-sampling and truncation techniques, the complexity of the architecture can be tailored to provide fast and low cost real-time motion estimation.

Methodology and Tools

The ability to (re)schedule a task either in hardware or software will be an important asset in a reconfigurable systems-on-chip. To support this feature Mignolet et al. have developed an infrastructure that, combined with a suitable design environment permits the implementation and management of hardware/software relocatable tasks. The authors present a strategy for heterogeneous context switching together with a description of the communication scheme and the design environment.

In Chapter 7, Wiangtong et al describe a semi-automatic co-design environment for the *UltraSONIC* reconfigurable computer. *UltraSONIC* is targeted at real-time video applications and consists of a single multitasking host processor augmented with multiple reconfigurable coprocessors. Tasks are assumed to be dataflow intensive with few control flow constructs. By treating hardware and software tasks uniformly, the authors are able to engineer a common run-time task manager with support for automatic partitioning and scheduling. They validate their flow by implementing an 8-point FFT algorithm.

In Chapter 8, Guo et al describe a design method to map applications written in a high level source language program, like C, to a coarse grain reconfigurable architecture, called MONTIUM. The source code is first translated into a control dataflow graph (CDFG). Then after applying graph clustering, scheduling and allocation on this CDFG, it can be mapped onto the target architecture. High performance and low power consumption are achieved by exploiting maximum parallelism and locality of reference respectively. Using a special mapping method, the flexibility of the MONTIUM architecture can be exploited.

The eXtreme Processing Platform (XPP) is a coarse-grained dynamically reconfigurable architecture which was already introduced in Chapter 3. Chapter 9 presents a compiler aiming to program the XPP using a subset of the C language. The compiler, apart from mapping the computational structures onto the available resources on the device, splits the program in temporal sections when it needs more resources than the physically available. In addition a scheme to split the program such that the reconfiguration overheads are minimized, taking advantage of the overlapping of the execution stages on different configurations is presented.

A rearrangement of the currently running functions is presented in Chapter 10 to obtain enough contiguous space to implement incoming functions, avoiding the spreading of their components and the resulting degradation of system performance. A novel active relocation procedure for Configurable Logic Blocks (CLBs) is presented, able to carry out online rearrangements and defragmenting the available FPGA resources without disturbing functions currently running.

Chapter 11 shows how the hardware parts of an embedded system can be implemented in a hardware byte code, which can be interpreted using a virtual hardware machine running on an arbitrary FPGA. The authors describe how this approach leads to fast, portable and reconfigurable designs, which run on any programmable target architecture.

Chapter 12 presents a new technique to improve the efficiency of data scheduling for multi-context reconfigurable architectures targeting multimedia and DSP applications. The main goal of this technique is to diminish application energy consumption. Two levels of on-chip data storage are assumed in the reconfigurable architecture. The authors show that a suitable data scheduling could decrease the energy required to implement the dynamic reconfiguration of the system.

Chapter 13 focuses on design methods for dynamic reconfiguration targeted to SoCs implemented on FPGAs. The main distinction between conventional digital systems design and the methods addressed here is the possibility of dynamically replacing hardware modules (i.e. IP cores) while the rest of the system is still working.

Applications

Chapter 14 describes an approach to hardware/software design space exploration for reconfigurable processors. The existing compiler tool-chain, because of the user-definable instructions, needs to be extended in order to offer developers an easy way to explore the design space. Such extension often is not easy to use for developers who have only a software background, and are not familiar with reconfigurable architecture details or hardware design. The authors illustrate their approach by using a real case study, a software implementation of a UMTS turbo-decoder that achieves a good speed-up of the whole application by exploiting user-defined instructions on the dynamically reconfigurable portion of the data path.

The work of Bellows et al. is motivated by the fact that network performance is doubling every eight months while the processing power of modern CPUs is failing to keep pace. Faced with handling higher network data rates and more complex encryption algorithms, the authors describe how they have developed an accelerator based on FPGAs called GRIP (gigabit-rate IPsec). GRIP is a network-processing accelerator that uses Xilinx FPGAs and integrates into a standard Linux-TCP/IP host equipped with IPsec, a set of protocols developed to support secure exchange of packets at the layer. The accelerator can handle full end-to-end, AES encrypted HDTV at 1 Gbps line rates on commodity networks. Meanwhile, Sourdis and Pnevmatikatos describe in Chapter 16 a network intrusion detection system (NIDS) designed to operate at the 10 Gbps OC192 line speeds. Using extensive parallelism and fine-grained pipelining, they have achieved throughput in excess of 11 Gbps with 50 rules from the open source Snort system.

In Chapter 17 Cilardo et al. present a modification of the popular Montgomery algorithm for modular multiplication through the addition of carry-save number representation. They evaluate their approach by implementing the Rivest-Shamir-Adleman (RSA) public-key cryptography algorithm. Their algorithm results in a highly modular, bit-sliced architecture that maps efficiently on to FPGAs and yields a 32% speed advantage over previously reported implementations.

The relative complexity of the discrete log problem in elliptic curve cryptography (ECC) can result in both increased speed and reduced key size for a given level of security. ECC has potential advantages in applications where bandwidth, power and silicon area are at a premium. Daly et al. propose a new architecture for modular division using what is essentially a look-ahead strategy. They pre-compute elements of the algorithm before finalizing the most time-consuming, magnitude comparison operation to yield circuits that are 40% faster than prior work. In Chapter 19, Kim and colleagues also focus on ECC. They propose a new systolic architecture that can perform both multiplication and division. This is a parameterizable architecture that is regular, modular and is unidirectional with respect to data flow. Consequently it can handle arbitrary irreducible poly-nomials and is scalable with respect to the size of the underlying Galois field.

McLoone and McCanny present an algorithm for decryption of the SHACAL-1 block cipher along with new implementations of the SHACAL-1/2 encryptionand decryption algorithms. SHACAL is a block cipher from Gemplus International that has its origins in ANSI's Secure Hashing Algorithm (SHA) used in encryption mode. The authors describe both compact, iterative imple-mentations and fast, highly-pipelined implementations. Impressive throughputs of 23 Gbps for SHACAL-1 and 26 Gbps for SHACAL-2 are reported with Xil-inx Virtex-II FPGAs.

Our discussion of security closes with a survey performed by Wollinger and Paar on the issues concerning the use of FPGAs in secure systems. The paper is noteworthy for the catalogue of methods of attack that it considers. Some of the more esoteric of these have not yet been applied to FPGAs to the best of the authors knowledge. While they do not recommend FPGAs in general for secure applications, the authors do note that physical attacks against FPGAs are very complex and even more difficult that analogous attacks against mask programmed ASICs.

Sousa and colleagues report on their use of FPGAs to develop a visual encoding system that may be used to construct a portable visual neuroprosthesis for profoundly blind people. The authors are able to exploit time-shared circuit techniques while still satisfying the real-time response of their system. They also exploit the FPGA reconfigurability to tailor devices to the requirements of individual patients.

In Chapter 23, Yu et al describe a new design for a Smith-Waterman systolic cell. The Smith-Waterman algorithm can be used for sequence alignment when searching large databases of DNA information. It makes all pair-wise compar-isons between string pairs thus achieving high sensitivity. However, this is offset by higher computation costs than more popular techniques, hence the motiva-tion for faster implementations. The authors design allows the edit distance between two strings to be computed without the need to use runtime reconfig-uration as was required with earlier implementations that achieved comparable performance.

In the final chapter, Lee et al. explore the influence of polynomial degrees on the calculation of piecewise approximation of mathematical functions using hierarchical segmentation. Mathematical functions can be computationally expensive to evaluate so approximations such as look-up tables are commonly deployed. However for complex functions, the size of the tables becomes impractical. An alternative is to segment the continuous function and perform piecewise approximation with one or more polynomials of varying degrees over the closed interval. Another innovation is to use a hierarchy of uniform segments and segments whose sizes vary by powers of two. Lee reports that hierarchical function segmentation (HFS) is particularly effective for non-linear regions and that it outperforms optimum uniform segmentation by a factor of two over a wide range of operand widths and polynomial degrees. Furthermore, second order polynomials start to outperform first order approximations at operand widths of sixteen bits.

About the Editors

Patrick Lysaght is a Senior Director in Xilinx Research Labs, San Jose, California where he is responsible for research into reconfigurable and embedded systems and for the Xilinx University Program. Prior to joining Xilinx in 2001, he was a Senior Lecturer in the Department of Electronic and Electrical Engineering at the University of Strathclyde in Glasgow and also at the Institute for System Level Integration in Livingston, Scotland. He has worked for Hewlett Packard in the UK in a number of roles including research and development, marketing, and sales. Patrick received his BSc (Electronic Systems) in 1981 from the University of Limerick, Ireland and his MSc (Digital Techniques) in 1987 from Heriot Watt University, Edinburgh, Scotland.

Patrick's research interests include real-time embedded systems, reconfigurable computing and system-level design. He is the author of more than forty technical papers and is active in the organization of several international conferences. He currently serves as chairman of the steering committee of the Field Programmable Logic Conference (FPL), the world's foremost conference dedicated to field programmable logic.

Wolfgang Rosenstiel is University Professor and holds the Chair for Computer Engineering at the University of Tübingen. He is also Managing Director of the Wilhelm Schickard Institute at Tübingen University, and Director for the Department for System Design in Microelectronics at the Computer Science Research Centre FZI. He is on the Executive Board of the German edacentrum.

His research areas include electronic design automation, embedded systems, computer architecture, multimedia, and artificial neural networks. He is member of the executive committees of DATE and ICCAD. As topic chair of the DATE program committee for configurable computing he was in charge of the contribution from DATE related papers to this book.

Acknowledgements

Several people deserve our most sincere gratitude for making this book possible. First and foremost among these are the many authors who have contributed their time and expertise to the project. The technical program committees of the FPL'03 and DATE'03 conferences also provided considerable help in the selection of the papers. We were guided by the expertise of their review panels in determining which papers to include.

We are grateful to the staff of Kluwer Academic Publishers for continuing to provide a path by which the latest research can be quickly brought to publication. Mark de Jongh, our commissioning editor, deserves special acknowledgement for being the champion of this project.

Patrick Lysaght would like to thank his wife, Janette, and his daughters, Emily, Rebecca and Stephanie, for all their love and support.

Wolfgang Rosenstiel would like to thank his wife, Christel, and his children, Alexander, Felix and Simone for their love and support.

Finally, we hope that the work compiled here will prove useful to existing researchers and inspire new ones to enter this exciting field of research for the first time.

Patrick Lysaght and Wolfgang Rosenstiel

July 2004

Architectures

Chapter 1

Extra-dimensional Island-Style FPGAs

Herman Schmit

Department of ECE
Carnegie Mellon University
Pittsburgh, PA, 15213 USA
herman_schmit@ieee.org

Abstract   This paper proposes modifications to standard island-style FPGAs that provide interconnect capable of scaling at the same rate as typical netlists, unlike traditionally tiled FPGAs. The proposal uses logical third and fourth dimensions to create increasing wire density for increasing logic capacity. The additional dimensions are mapped to standard two-dimensional silicon. This innovation will increase the longevity of a given cell architecture, and reduce the cost of hardware, CAD tool and Intellectual Property (IP) redesign. In addition, extra-dimensional FPGA architectures provide a conceptual unification of standard FPGAs and time-multiplexed FPGAs.

Keywords: FPGA Architecture, Interconnect, Rent Exponent

Introduction

Island-style FPGAs consist of mesh interconnection of logic blocks, connection blocks and switchboxes. Commercial FPGAs from vendors such as Xilinx are implemented in an island-style. One advantage of island-style FPGAs is that they are completely tileable. A single optimized hardware tile design can be used in multiple products all with different capacity, pin count, package, etc., allowing FPGA vendors to build a whole product line out of a single relatively small design. The tile also simplifies the CAD tool development effort. The phases of technology mapping, placement, routing and configuration generation are all simplified by the regularity and homogeneity of the island-style FPGA. Finally, these tiled architectures are amenable to a re-use based design methodology. A design created for embedding in larger designs, often called a

P. Lysaght and W. Rosenstiel (eds.),
New Algorithms, Architectures and Applications for Reconfigurable Computing, 3–13.
© 2005 *Springer. Printed in the Netherlands.*

core, can be placed to any location within the array, while preserving routing and timing characteristics.

The inherent problem with tiled architectures is that they do not scale to provide the interconnect typical of digital circuits. The relationship of design interconnect and design size is called Rent's rule. This relationship is described in [12], where it is shown that a partition containing n elements has Cn^p pins either entering or leaving the partition. The quantity p, which is known as the Rent exponent, has been empirically determined to be in the range from 0.57 to 0.75 in a variety of logic designs. In a tiled architecture, the number of logic elements in any square bounding box is proportional to the number of tiles contained in the square, and the number of wires crossing the square is proportional to the perimeter of the square. Therefore, in a traditional tiled architecture, the interconnect is a function of the square root of the logic capacity, and the "supplied" Rent exponent is $1/2$. Supply therefore can never keep up with interconnect demand, and at some point a tiled architecture will fail to have enough interconnect.

Commercial vendors have recognized this phenomenon. Succeeding generations of FPGAs always have more interconnect per logic block. This has the effect of keeping up with the Rent exponent by increasing the constant interconnect associated with each tile, i.e. increasing the C term in Rent's formula. There are several problems with this solution. First, it requires that the tile must be periodically redesigned, and all the CAD tools developed for that tile must be updated, tested and optimized. Second, all the cores designed for an early architecture cannot be trivially mapped to the new architecture.

What would be preferable would be to somehow scale the amount of interconnect between cells, while still providing the benefits of a tiled architecture, such as reduced hardware redesign cost and CAD tool development. This is impossible in conventional island-style FPGAs because they are two-dimensional, which means that the Rent exponent of supplied interconnect is fixed at one half. This paper proposes a new island-style architecture, based on three- and four-dimensional tiling of blocks, that can support Rent exponents greater than 0.5 across an entire family of FPGAs.

Three-dimensional FPGAs have been proposed frequently in the literature [14], [13], [1], [2], [8], but commercial three-dimensional fabrication techniques, where the transistors are present in multiple planes, do not exist. An optoelectrical coupling for three-dimensional FPGAs has also been proposed in [6], but there are no instances of such couplings in large commercial designs as yet. Four-dimensional FPGAs face even greater challenges to actual implementation, at least in this universe. Therefore, this paper proposes ways to implement three- and four-dimensional FPGAs in planar silicon technology. Surprisingly, these two-dimensional interconnect structures resemble double and quad interconnect lines provided in commercial island-style FPGAs, albeit with different scaling behavior.

1.1 Architecture

Extra-dimensional FPGAs can provide interconnect that matches the interconnect required by commercial designs and that scales with design size. This section describes how these extra dimensions provide scalable interconnect. Possible architectures for multi-dimensional FPGAs are discussed in the context of implementation on two-dimensional silicon. Three- and four-dimensional FPGA architectures are proposed, which will be the subject of placement and routing experiments performed in subsequent sections.

As discussed previously, a 2-D tiled FPGA provides a Rent exponent of 1/2 because of the relationship of the area and perimeter of a square. In a three-dimensional FPGA, assuming that all dimensions grow at an equal rate, the relationship between volume and surface area is governed by an exponent of 2/3. Therefore in a perfect three-dimensional FPGA, the supplied interconnect has a Rent exponent of 2/3. By extension, the Rent exponent for a four-dimensional architecture is 3/4.

There are many ways to construct an extra-dimensional FPGA. A conventional island-style FPGA can be envisioned as an array of CLBs, each of which is surrounded by a square of interconnect resources. The CLB connects to each of the edges of the square that surrounds it, as shown in Figure 1.1(a). By extension, a 3-D FPGA would have a three dimensional interconnection mesh, with CLBs located in the center of the cubes that make up this mesh. Each CLB would

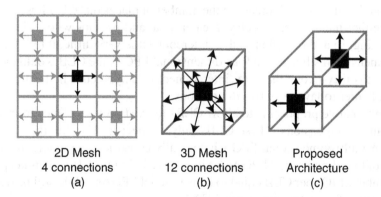

2D Mesh	3D Mesh	Proposed
4 connections	12 connections	Architecture
(a)	(b)	(c)

Figure 1.1. Two and three dimensional interconnect points: as shown in (a) a CLB in a 2D mesh must connect to 4 wires. A CLB in a 3D mesh must (b) connect to twelve points. The proposed architecture has 2D CLBs interconnected to the squares in a 3D mesh.

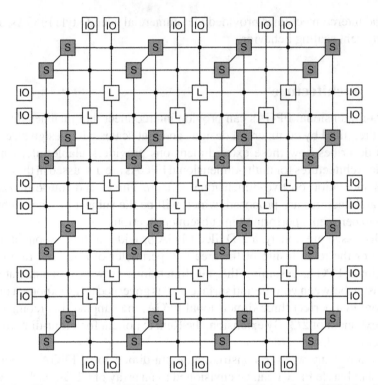

Figure 1.2. Three dimensional FPGA in two dimensions: This particular FPGA is a 3 × 3 × 2 array of CLBs. There are IOs around the periphery. Switch Blocks are labelled "S" and CLBs are labelled "L". Connections between xy planes take place in the diagonal channels between switch boxes. All lines in this figure are channels containing multiple wires.

have to connect to all twelve edges of the cube, as illustrated in Figure 1.1(b). This entails a three-fold increase in the number of places that CLB IOs connect, which either means greater delay, greater area, and perhaps worse placement and routing. In addition, this "cube" interconnect is very difficult to lay out in two-dimensional silicon. In a four-dimensional FPGA, the CLB would have to connect to thirty-two of the wires interconnecting a hypercube in the mesh, and is even harder to lay out in two dimensions.

An alternative proposal is to allow CLBs to exist only on planes formed by the x and y dimensions. These CLBs only connect to wires on the same xy plane. A multi-dimensional FPGA is logically constructed by interconnecting planes of two-dimensional FPGAs, as illustrated in Figure 1.1(c). This keeps the the number of IOs per CLB equal to conventional FPGAs, while still providing the benefits of an extra-dimensional FPGA.

In our proposed architecture, two adjacent xy planes of CLBs are interconnected by the corresponding switch boxes. For example, in a logically three-dimensional FPGA, the switch box at (x, y) in plane z is connected to the

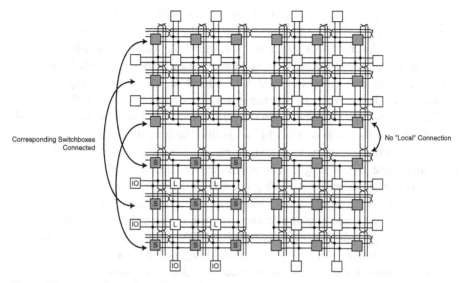

Figure 1.3. Four dimensional FPGA: This FPGA has dimensions of $2 \times 2 \times 2 \times 2$ CLBs. The connections between dimensions take place on the long lines that go from one switch box to the corresponding box in another dimension.

switch box at (x, y) in plane $z + 1$ and plane $z - 1$. Such a three-dimensional switchbox architecture was used in [1].

A problem with these extra-dimensional switch boxes is the increased number of transistors necessary to interconnect any wire coming from three or four dimensions. When four wires meet within a switch box, coming from all four directions in a two-dimensional plane, six transistors are necessary to provide any interconnection. With three dimensions, six wires come from every direction, requiring 15 transistors to provide the same flexibility of interconnect. With four dimensions, 28 transistors are necessary at each of these switch points. We will assume that extra-dimensional switch boxes include these extra number of transistors. We will later show that this overhead is mitigated by the reduction in channel width due to extra-dimensional interconnect. It is worth noting that there are only 58 possible configurations of switch point in a 3D switch box,[1] and 248 possible configurations of a 4D switch point. Therefore the amount of configuration for these switch points could potentially be significantly reduced to six or eight bits. It is also possible that all 15 or 28 transistors are not necessary to provide adequate interconnect flexibility. Future work will explore optimizations of the switch box architectures for extra-dimensional FPGAs.

Figure 1.2 and Figure 1.3 show feasible layouts for three- and four-dimensional FPGAs in two-dimensional silicon. The four-dimensional layout is particularly interesting, because it allows for close interconnections between neighboring CLBs, and because it is symmetric in both x and y dimensions. The most interesting aspect of the four-dimensional FPGA is how the interconnect resembles the interconnect structure of commercial island-style FPGAs. Particularly, the interconnections between different xy planes resemble the double and quad lines in Xilinx 4000 and Virtex FPGAs. These lines skip across a number of CLBs, in order to provide faster interconnect over long distance.

The primary difference between the traditional island-style with double and quad lines and the proposed 4D FPGA is that the length of the long wires increases when the size of the xy plane increases. This is the key to providing scalable interconnect. The length of wires in the two-dimensional implementation of the 4D FPGA increases as the logical size of the device increases. This does mean that the delay across those long wires is a function of the device size, which presents an added complication to timing-oriented design tools. Fortunately, interconnection delay in commercial FPGAs is still dominated by the number of switches in any path and not the length of the net.[2]

The assumption that the physical channel width is proportional to the number of wires passing through that channel leads to the conclusion that the four-dimensional FPGA in Figure 1.3 would be much larger than the two-dimensional FPGA with the same number of CLBs. In real FPGA layouts however, the largest contributor to channel width is the number of programmable interconnect points in the channel, and the width of the switch box. Because these points require one to six transistors, they are much larger than the minimum metal pitch. Most wires in the channel of the four-dimensional FPGA do not contact any interconnect points however. These wires can be tightly routed at a higher level of metal, and may only minimally contribute to total channel width.

A second substantial difference between the proposed 4D FPGA and the traditional island-style FPGA is that in the commercial FPGA, the local connections are never interrupted. In the 4D FPGA, there is no local interconnection from the boundary of one xy plane to the corresponding boundary of the adjacent xy plane. The reason for this is to provide for tiling of IP blocks. Suppose we want to use a four-dimensional core in a four-dimensional FPGA. To guarantee that this core will fit in this FPGA, all four dimensions of the core must be smaller than the four dimensions of the FPGA. If there are local connections between xy planes, that cannot be guaranteed. A core that was built on a device with $x = 3$, and which used local connections between two xy planes would only fit onto other devices with $x = 3$.

The next section will describe an experiment that compares the scaling of interconnect in two- and four-dimensional FPGA. First, Rent's rule is demonstrated by measuring how the channel width of a two-dimensional FPGA increases as the size of the design increases. When the same suite of netlists is placed and routed on an 4D FPGA, the channel width remains nearly constant.

1.2 Experimental Evaluation

In order to demonstrate the effectiveness of extra-dimensional FPGAs, a large set of netlists have been placed and routed on both two- and four-dimensional FPGAs. The minimum channel width required in order to route these designs is compared. In the two-dimensional FPGA, the channel width grows with respect to the logical size of the design. The four-dimensional FPGA scales to provide exactly the required amount of interconnect, and channel width remains nearly constant regardless of the design size.

A significant obstacle to performing this experiment was to acquire a sufficiently large set of netlists in order to make conclusions based on measured data. Publicly available circuit benchmark suites are small, both in the number of gates in the designs, and the number of designs in the suites. In this paper we use synthetic netlists generated by the CIRC and GEN tools [10, 11]. CIRC measures graph characteristics, and has been run on large sets of commercial circuits. GEN constructs random netlists that have the same characteristics as the measured designs. We used the characteristics measured by CIRC on finite state machines to generate a suite of 820 netlists ranging from 20 to 585 CLBs.

The placement and routing software we will use is an extended version of the VPR tool [3]. The modifications to this software were relatively simple. The placement algorithm is extended to use a multi-dimensional placement cost. The cost function is a straight-forward four-dimensional extension of the two-dimensional cost, which is based on the RISA cost function [4]. The cost of routing in the two additional dimensions is scaled by a small factor to encourage nets to be routed in a single xy plane.

The logic block architecture used for this experiment is the standard architecture used in VPR. This architecture consist of a four-input lookup table (LUT) and a bypassable register. The switch box architecture is the Xilinx-type (also known as subset or disjoint), where the n-th channel from one direction connects to the n-th channel in every other direction.

VPR works by first placing the design, and then attempting to find the minimal channel width that will allow routing of the design. It performs a binary search in order to find this minimal channel width. The results of routing the netlist suite on the two-dimensional and four-dimensional architecture are shown in Figure 1.4, which plots the maximum channel width versus the netlist size in CLBs. The best-fit power graph has been fit to this data, showing an exponent of 0.28 for the two-dimensional FPGA. As shown in [7] and [9], this indicates the Rent exponent of these designs is approximately $0.28 + 0.5 = 0.78$, which is large, but possibly reasonable for the design of an FPGA architecture.

Even a four dimensional FPGA cannot provide interconnect with a Rent exponent of 0.78. In our experiment, the extra dimensions of the FPGA scale at one-eight the rate of the x and y dimensions. If the x and y dimensions are less

10

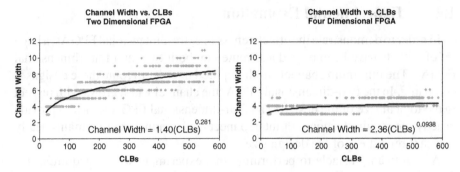

Figure 1.4. Two- and Four-Dimensional FPGA Channel Width.

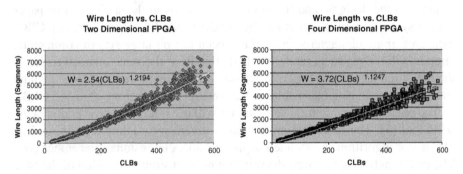

Figure 1.5. Wirelengths for Two- and Four-dimensional FPGAs.

than eight, there is no other extra dimensions to the FPGA (the FPGA is simply two-dimensional). With x or y in the range of 8 to 16, there are two planes of CLBs in each of the extra dimensions.

The channel width for the four-dimensional FPGA, as shown in Figure 1.4 is comparatively flat across the suite of netlists. Figure 1.5 demonstrates the effectiveness of the technique by showing total wire length, in channel segments, for each design. When fit to a power curve, the two dimensional FPGA exhibits a much larger growth rate than the four dimensional FPGA. Superlinear growth of wire length was predicted by [7] as a result of Rent's rule. The four dimensional FPGA evidences a nearly linear relationship between wire length and CLBs, indicating that the supply of interconnect is keeping pace with the demand.

As mentioned in previously, the four dimensional switch box might have more than four times more bits per channel compared to the two dimensional switch box. Table 1.1 shows the number of configuration bits necessary to program the interconnect for a two- and four-dimensional FPGA with 576 CLBs. Using our data from the previous experiments, we have assumed a channel width of twelve for the two-dimensional FPGA, and six for the

Table 1.1. Computation of configuration bits required for large two- and four- dimensional FPGAs. The larger number of bits per switch point the four-dimensional FPGA is countered by the reduced channel width. If the switch point configuration is encoded with sixteen bits, there is no configuration overhead.

FPGA Dimensions	Channel Width	Switch Point Bits	Total Switchbox Bits	Total Channel Bits	Total Config Bits
24 × 24	12	6	41,472	27,648	69,120
2 × 12 × 12 × 2	6	28	96,768	13,824	110,592
2 × 12 × 12 × 2	6	16	55,296	13,824	69,120

four-dimensional FPGA. Considering both switch box and connection box bits, the four-dimensional FPGA requires 60% more configuration bits. If we can control the four-dimensional switch point with just sixteen bits, then the reduction in channel width completely compensates for the more complex switch point.

Four-dimensional FPGAs provide scalable interconnect across a wide suite of netlists. Required channel width can remain constant across a large family of devices, allowing the hardware and CAD tool development efforts to be amortized over many devices, and reducing the waste of silicon in smaller devices.

1.3   Time Multiplexing and Forward-compatiblity

Since their proposal [17, 5, 15], time-multiplexed FPGAs have been thought of as fundamentally different creatures from standard FPGAs. The Xilinx and Sanders FPGAs operate by separating every logical cycle into multiple micro-cycles. Results from a micro-cycle are passed to subsequent micro-cycles through registers.

Time-multiplexed FPGAs could also be constructed using a three-dimensional FPGA architecture like that shown in Figure 1.2. Micro-registers would be inserted on the inter-plane connections that exist between switchboxes. The number of registers between planes would correspond to the channel width, which allows it to be designed to provide scalable interconnect.

By viewing time as a third dimension, the scheduling task can be accomplished within the scope of a three dimensional placement algorithm. All combinational paths must go from a previous configuration, or plane, to a future one, and not vice-versa. Therefore real logical registers must be placed in a configuration that is evaluated after its entire fan-in cone. This is conceptually much simpler than the separate scheduling and place-and-route phases implemented in [16].

1.4   Conclusions

This paper discussed the benefits of three- and four-dimensional FPGAs, and their implementation in conventional two-dimensional silicon. Primary among the benefits is the ability to provide interconnect that scales at the same rate as typical netlists. Four dimensional FPGAs resemble the double and quad lines in commercial FPGAs, although the key to providing scalable interconnect is to increase the length of those lines as the device grows.

Notes

1. This is determined by $1 + \binom{6}{2} + \binom{6}{3} + \binom{6}{4} + \binom{6}{5} + \binom{6}{6} = 58$.
2. The increasing "logical length" of these lines may also be counter-acted to some extent by the scaling of technology.

References

[1] M. J. Alexander, J. P. Cohoon, J. L. Colflesh, J. Karro, and G. Robins. Three-dimensional field-programmable gate arrays. In *Proceedings of the IEEE International ASIC Conference*, pages 253–256, September 1995.

[2] M. J. Alexander, J. P. Cohoon, J. Karro J. L. Colflesh, E. L. Peters, and G. Robins. Placement and routing for three-dimensional FPGAs. In *Fourth Canadian Workshop on Field-Programmable Devices*, pages 11–18, Toronto, Canada, May 1996.

[3] V. Betz and J. Rose. Effect of the prefabricated routing track distribution on FPGA area efficiency. *IEEE Transactions on VLSI Systems*, 6(3):445–456, September 1998.

[4] C. E. Cheng. RISA: Accurate and efficient placement routability modeling. In *Proceedings of IEEE/ACM International Conference on CAD (ICCAD)*, pages 690–695, November 1996.

[5] A. DeHon. DPGA-coupled microprocessors: Commodity ICs for the early 21st century. In D. A. Buell and K. L. Pocek, editors, *Proceedings of IEEE Workshop on FPGAs for Custom Computing Machines*, pages 31–39, Napa, CA, April 1994.

[6] J. Depreitere, H. Neefs, H. Van Marck, J. Van Campenhout, R. Baets, B. Dhoedt, H. Thienpont, and I. Veretennicoff. An optoelectronic 3-D field programmable gate array. In R. Hartenstein and M. Z. Servit, editors, *Field-Programmable Logic: Architectures, Synthesis and Applications. 4th International Workshop on Field-Programmable Logic and Applications*, pages 352–360, Prague, Czech Republic, September 1994. Springer-Verlag.

[7] W. E. Donath. Placement and average interconnection lengths of computer logic. *IEEE Transactions on Circuits and Systems*, pages 272–277, April 1979.

[8] Hongbing Fan, Jiping Liu, and Yu-Liang Wu. General models for optimum arbitrary-dimension fpga switch box designs. In *Proceedings of IEEE/ACM International Conference on CAD (ICCAD)*, pages 93–98, November 2000.

[9] A. El Gamal. Two-dimensional stochastic model for interconnections in master slice integrated circuits. *IEEE Transactions on Circuits and Systems*, 28(2):127–138, February 1981.

[10] M. Hutton, J.P. Grossman, J. Rose, and D. Corneil. Characterization and parameterized random generation of digital circuits. In *Proceedings of the 33rd ACM/SIGDA Design Automation Conference (DAC)*, pages 94–99, Las Vegas, NV, June 1996.

[11] M. Hutton, J. Rose, and D. Corneil. Generation of synthetic sequential benchmark circuits. In *5th ACM/SIGDA International Symposium on Field Programmable Gate Arrays, (FPGA 97)*, February 1997.

[12] B. S. Landman and R. L. Russo. On pin versus block relationship for partions of logic circuits. *IEEE Transactions on Computers*, C-20:1469–1479, 1971.

[13] M. Leeser, W. M. Meleis, M. M. Vai, W. Xu S. Chiricescu, and P. M. Zavracky. Rothko: A three-dimensional FPGA. *IEEE Design and Test of Computers*, 15(1):16–23, January 1998.

[14] W. M. Meleis, M. Leeser, P. Zavracky, and M. M. Vai. Architectural design of a three dimensional FPGA. In *Proceedings of the 17th Conference on Advanced Research in VLSI (ARVLSI)*, pages 256–268, September 1997.

[15] S. Scalera and J. R. Vazquez. The design and implementation of a context switching FPGA. In D. A. Buell and K. L. Pocek, editors, *Proceedings of IEEE Workshop on FPGAs for Custom Computing Machines*, pages 78–85, Napa, CA, April 1998.

[16] S. Trimberger. Scheduling designs into a time-multiplexed FPGA. In *6th ACM/SIGDA International Symposium on Field Programmable Gate Arrays, (FPGA 98)*, pages 153–160, February 1998.

[17] S. Trimberger, D. Carberry, A. Johnson, and J. Wong. A time-multiplexed FPGA. In J. Arnold and K. L. Pocek, editors, *Proceedings of IEEE Workshop on FPGAs for Custom Computing Machines*, pages 22–28, Napa, CA, April 1997.

Chapter 2

A Tightly Coupled VLIW/Reconfigurable Matrix and its Modulo Scheduling Technique

Bingfeng Mei,[1,2] Serge Vernalde,[2] Diederik Verkest,[1,2,3] and Rudy Lauwereins[1,2]

[1] *IMEC vzw, Kalpeldreef 75, Leuven, Belgium*

[2] *Department of Electrical Engineering, Katholieke Universiteit Leuven, Belgium*

[3] *Department of Electrical Engineering, Vrije Universiteit Brussel, Belgium*

2.1 Introduction

Coarse-grained reconfigurable architectures have become increasingly important in recent years. Various architectures have been proposed [1–4]. These architectures often comprise a matrix of functional units (FUs), which are capable of executing word- or subword-level operations instead of bit-level ones found in common FPGAs. This *coarse* granularity greatly reduces the delay, area, power and configuration time compared with FPGAs. However, these advantages are achieved at the expense of flexibility. Usually the reconfigurable matrix alone is not able to execute entire applications. Most coarse-grained architectures are coupled with processors, typically RISCs. The computational-intensive kernels, typically loops, are mapped to the matrix, whereas the remaining code is executed by the processor. So far not much attention has been paid to the integration of these two parts. The coupling between the processor and the matrix is often loose, consisting essentially of two separate parts connected by a communication channel. This results in programming difficulty and communication overhead. In addition, the coarse-grained reconfigurable architecture consists of components which are similar to those used in processors. This resource-sharing opportunity is not extensively exploited in traditional coarse-grained architectures.

15

P. Lysaght and W. Rosenstiel (eds.),
New Algorithms, Architectures and Applications for Reconfigurable Computing, 15–28.
© 2005 *Springer. Printed in the Netherlands.*

To address these problems, in this chapter we present an architecture called ADRES (*Architecture for Dynamically Reconfigurable Embedded System*), which tightly couples a VLIW (very long instruction word) processor and a coarse-grained reconfigurable matrix. The VLIW processor and the coarse-grained reconfigurable matrix are integrated into one single architecture but with two virtual functional views. This level of integration has many advantages compared with other coarse-grained architectures, including improved performance, a simplified programming model, reduced communication costs and substantial resource sharing.

Any new architecture can hardly be successful without good design methodology. Therefore, we developed a compiler for ADRES. The central technology is a novel modulo scheduling technique, which is able to map kernel loops to the reconfigurable matrix by solving placement, routing and scheduling simultaneously in a modulo constrained space. Combined with traditional ILP (instruction-level parallelism) compilation techniques, our compiler can map an entire application to the ADRES architecture efficiently and automatically.

This chapter is organized as follows: section 2.2 discusses the architectural aspects of ADRES; section 2.3 describes the modulo scheduling algorithm in details; section 2.4 presents the experimental results and section 2.5 concludes the chapter.

2.2   ADRES Architecture

2.2.1   Architecture Description

The ADRES architecture (Fig 2.1) consists of many basic components, including mainly FUs and register files(RFs), which are connected in a certain topology. The FUs are capable of executing word-level operations selected by a control signal. The RFs can store intermediate data. The whole ADRES matrix has two functional views, the VLIW processor and the reconfigurable matrix. These two functional views share some physical resources because their executions will never overlap with each other thanks to the processor/co-processor model. For the VLIW processor, several FUs are allocated and connected together through one multi-port register file, which is typical for VLIW architecture. Some of these FUs are connected to the memory hierarchy, depending on available ports. Thus the data access to the memory is done through the load/store operations available on those FUs.

For the reconfigurable matrix part, apart from the FUs and RF shared with the VLIW processor, there are a number of *reconfigurable cells* (RC) which basically comprise FUs and RFs too (Fig. 2.2). The FUs can be heterogeneous supporting different operation sets. To remove the control flow inside loops,

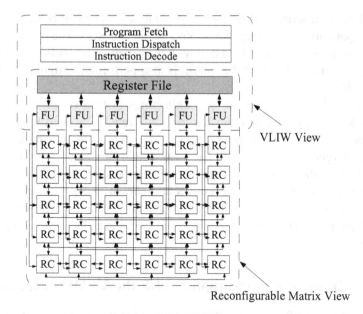

Figure 2.1. ADRES core

the FUs support predicated operations. The distributed RFs are small with less ports. The multiplexors are used to direct data from different sources. The configuration RAM stores a few configurations locally, which can be loaded on cycle-by-cycle basis. The configurations can also be loaded from the memory hierarchy at the cost of extra delay if the local configuration RAM is not big

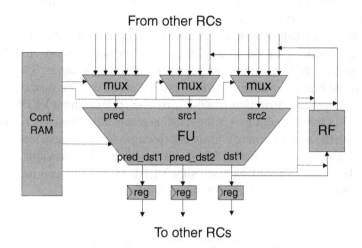

Figure 2.2. Example of a Reconfigurable Cell

enough. Like instructions in ISPs, the configurations control the behaviour of the basic components by selecting operations and multiplexors. The reconfigurable matrix is used to accelerate the dataflow-like kernels in a highly parallel way. The access to the memory of the matrix is also performed through the VLIW processor FUs.

In fact, ADRES is a template of architectures instead of a fixed architecture. An XML-based architecture description language is used to define the communication topology, supported operation set, resource allocation and timing of the target architecture [5]. Even the actual organization of the RC is not fixed. FUs and RFs can be put together in several ways. For example, two FUs can share one RF. The architecture shown in Fig. 2.1 and Fig. 2.2 is just one possible instance of the template. The specified architecture will be translated to an internal architecture representation to facilitate compilation techniques.

2.2.2   Improved Performance with the VLIW Processor

Many coarse-grained architectures consist of a reconfigurable matrix and a relatively slow RISC processor, e.g. TinyRisc in MorphoSys [1] and ARM in PACT XPP [3]. These RISC processors execute the unaccelerated part of the application, which only represents a small portion of execution time. However, such a system architecture has problems due to the huge performance gap between the RISC and the matrix. According to Amdahl's law [6], the performance gain that can be obtained by improving some portion of an application can be calculated according to equation 2.1. Suppose the kernels, representing 90% of execution time, are mapped to the reconfigurable matrix to obtain an acceleration of 30 times over the RISC processor, the overall speedup is merely 7.69. Obviously a high kernel speedup is not translated to a high overall speedup. Speeding up the unaccelerated part, which is often irregular and control-intensive code, is important for the overall performance. Though it is hard to exploit higher parallelism from it on the reconfigurable matrix, it is still possible to discover *instruction-level parallelism* (ILP) using a VLIW processor, where 2–4 times speedup over the RISC is reasonable. If we recalculate the speedup with the assumption of 3 times acceleration for the unaccelerated code, the overall acceleration is now 15.8, much better than the previous scenario. This simple calculation proves that using VLIW in ADRES can improve the overall speedup dramatically in certain circumstances.

$$Speedup_{overall} = \frac{1}{(1 - Fraction_{enhanced}) + \frac{Fraction_{enhanced}}{Speedup_{enhanced}}} \qquad (2.1)$$

2.2.3   Simplified Programming Model and Reduced Communication Cost

A simplified programming model and reduced communication cost are two important advantages of the ADRES architecture. These are achieved by making the VLIW processor and the reconfigurable matrix share access to the memory and the register file.

In other reconfigurable architectures, the processor and the matrix are essentially separated. The communication is often through explicit data copying. Though some techniques are adopted to reduce the data copying, e.g., wider data bus and DMA controller, the overhead is still considerable in terms of performance and energy. From the programming point of view, the separated processor and reconfigurable matrix require significant code rewriting. Starting from a software implementation, we have to identify the data structures used for communication and replace them with communication primitives. Data analysis should be done to make sure as few as possible data are actually copied. In addition, the kernels and the remaining code have to be cleanly separated in such a way that no shared access to any data structure remains. These transformations are often complex and error-prone.

In ADRES, the data communication is performed through the shared RF and memory. This feature is very helpful to map high-level language code such as C without major changes. When a high-level language is compiled to a processor, the local variables are allocated in the RF, while the static variables and arrays are allocated in the memory space. When the control of the program is transferred between the VLIW processor and the reconfigurable matrix, those variables used for communication can stay in the RF or the memory as they were. The copying is unnecessary because both the VLIW and the reconfigurable matrix share access to the RF and memory hierarchy. The code doesn't require any rewriting and can be handled by the compiler automatically.

2.2.4   Resource Sharing

Since the basic components such as the FUs and RFs of the reconfigurable matrix and those of the VLIW processor are basically the same, it is natural to think that resources might be shared to have substantial cost-saving. In other coarse-grained architectures, the resources cannot be effectively shared because the processor and the matrix are two separate parts. For example, the FU in the TinyRisc of MorphoSys cannot work cooperatively with the reconfigurable cells in the matrix. In ADRES, since the VLIW processor and the reconfigurable matrix are indeed two virtual functional views of the same physical entity, many resources are shared among these two parts. Due to its processor/co-processor model, only the VLIW processor or the reconfigurable matrix is active at any time. This fact makes resource sharing possible. Resource sharing of the powerful FUs and the multi-port RF of the VLIW by the matrix can greatly improve the performance and schedulability of kernels mapped on the matrix.

2.3   Modulo Scheduling

The objective of modulo scheduling is to engineer a schedule for one iteration of the loop such that this same schedule is repeated at regular intervals with respect to intra- and inter-iteration dependency and resource constraints. This interval is termed the *initiation interval* (II), essentially reflecting the performance of the scheduled loop. Various effective heuristics have been developed to solve this problem for both unified and clustered VLIWs [9, 11–13]. However, they cannot be applied to a coarse-grained reconfigurable architecture because the nature of the problem becomes more difficult, as illustrated next.

2.3.1   Problem Illustrated

To illustrate the problem, consider a simple dependency graph, representing a loop body, in Fig. 2.3a and a 2×2 matrix in Fig. 2.3b. The scheduled loop is depicted in Fig. 2.4a, where the 2×2 matrix is flattened to 1×4 for convenience of drawing. Nevertheless, the topology remains the same.

Fig 2.4a is a space-time representation of the scheduling space. From Fig. 2.4a, we see that modulo scheduling on coarse-grained architectures is a combination of 3 sub-problems: *placement, routing and scheduling*. Placement determines on which FU of a 2D matrix to place one operation. Scheduling, in its literal meaning, determines in which cycle to execute that operation. Routing connects the placed and scheduled operations according to their data dependencies. If we view *time* as an axis of 3D space, the modulo scheduling can be simplified to a placement and routing problem in a modulo-constrained 3D space, where the routing resources are asymmetric because any data can only

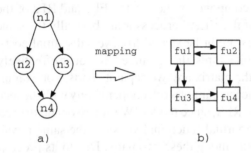

Figure 2.3.   a) A simple dataflow graph; b) A 2×2 reconfigurable matrix

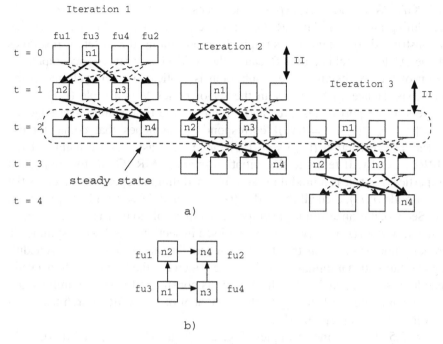

Figure 2.4.   a) Modulo scheduling example; b) Configuration for 2×2 matrix

be routed from smaller time to bigger time, as shown in Fig. 2.4a. Moreover, all resources are modulo-constrained because the execution of consecutive iterations which are in distinct stages is overlapped. The number of stages in one iteration is termed *stage count* (SC). In this example, II $= 1$ and SC $= 3$. The schedule on the 2×2 matrix is shown in Fig. 2.4b. FU1 to FU4 are configured to execute n2, n4, n1 and n3 respectively. In this example, there is only one configuration. In general, the number of configurations that need to be loaded cyclically is equal to II.

By overlapping different iterations of a loop, we are able to exploit a higher degree of ILP. In this simple example, the instruction per cycle (IPC) is 4. As a comparison, it takes 3 cycles to execute one iteration in a non-pipelined schedule due to the data dependencies, corresponding to an IPC of 1.33, no matter how many FUs in the matrix.

2.3.2   Modulo Routing Resource Graph

We develop a graph representation, namely *modulo routing resource graph* (MRRG), to model the ADRES architecture internally for the modulo scheduling algorithm. MRRG combines features of the *modulo reservation table*(MRT) [7] for software pipelining and the *routing resource graph* [8] used in FPGA P&R, and only exposes the necessary information to the modulo scheduling algorithm. An MRRG is a directed graph $G = \{V, E, II\}$ which is constructed by composing sub-graphs representing the different resources of the ADRES architecture. Because the MRRG is a time-space representation of the architecture, every subgraph is replicated each cycle along the time axis. Hence each node v in the set of nodes V is a tuple (r, t) where r refers to the port of resource and t refers to the time stamp. The edge set $E = \{(v_m, v_n)|t(v_m) <= t(v_n)\}$ corresponds to switches that connect these nodes. The restriction $t(v_m) <= t(v_n)$ models the asymmetric nature of the MRRG. Finally, II refers to the initiation interval. MRRG has two important properties. First, it is a modulo graph. If scheduling an operation involves the use of node (r, t_j), then all the nodes $\{(r, t_k)|t_j \bmod II = t_k \bmod II\}$ are used too. Second, it is an asymmetric graph. It is impossible to find a route from node v_i to v_j, where $t(v_i) > t(v_j)$. As we will see in section 2.3.3, this asymmetric property imposes big constraints on the scheduling algorithm. During scheduling we start with a minimal II and iteratively increase the II until we find a valid schedule (see section 2.3.3). The MRRG is constructed from the architecture specification and the II to try. Each component of the ADRES architecture is converted to a subgraph in MRRG.

Fig. 2.5 shows some examples. Fig. 2.5a is a 2D view of a MRRG subgraph corresponding to a FU, which means in the real MRRG graph with time dimension, all the subgraphs have to be replicated each cycle along the time

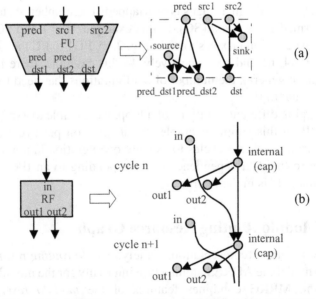

Figure 2.5. MRRG representation of ADRES architecture parts

axis. For FUs, all the input and output ports have corresponding nodes in the MRRG graph. Virtual edges are created between src1 and dst, src2 and dst, etc. to model the fact that a FU can be used as routing resource to directly connect src1 or src2 to dst, acting just like a multiplexor or demultiplexor. In addition, two types of artificial nodes are created, namely *source* and *sink*. When a commutative operation, e.g., *add*, is scheduled on this FU, the source or sink node are used as routing terminals instead of the nodes representing ports. Thus the router can freely choose which port to use. This technique improves the flexibility of the routing algorithm and leads to higher routability. Fig. 2.5b shows a space-time MRRG subgraph for a register file with one write port and two read ports. The idea is partly from [10]. Similar to the FU, the subgraph has nodes corresponding to each input and output port, which are replicated over each cycle. Additionally, an internal node is created to represent the capacity of the register file. All internal nodes along the time axis are connected one by one. The input nodes are connected to the internal node of next cycle, whereas the output nodes are connected to the internal node of this cycle. In this way, the routing capability of the register file is effectively modelled via its write-store-read functionality. Moreover, a *cap* property is associated with the internal node which is equal to the capacity of the register file. Therefore, the register allocation problem is implicitly solved by our scheduling algorithm. Other types of components such as bus and multiplexor can be modelled in a similar way. This abstraction, all routing resources, whether physical or virtual, are modelled in a universal way using nodes and edges. This unified abstract view of the architecture exposes only necessary information to the scheduler and greatly reduces the complexity of the scheduling algorithm.

2.3.3   Modulo Scheduling Algorithm

By using MRRG, the three sub-problems (placement, routing and scheduling) are reduced to two sub-problems (placement and routing), and modulo constraints are enforced automatically. However, it is still more complex than traditional FPGA P&R problems due to the modulo and asymmetric nature of the P&R space and scarcity of available routing resources. In FPGA P&R algorithms, we can comfortably run the placement algorithm first by minimizing a good cost function that measures the quality of placement. After minimal cost is reached, the routing algorithm connects placed nodes. The coupling between these two sub-problems is very loose. In our case, we can hardly separate placement and routing as two independent problems. It is very hard to find a placement algorithm and a cost function which can foresee the routability during the routing phase. Therefore, we propose a novel approach to solve these two sub-problems in one framework. The algorithm is described in Fig. 2.6.

```
SortOps();
II   := MII(DDG);

while not scheduled do
  InitMrrg(II);
  InitTemperature();
  InitPlaceAndRoute();          (1)

  while not scheduled do
    for each op in sorted operation list
      RipUpOp();

      for i := 1 to random_pos_to_try do
        pos := GenRandomPos();
        success := PlaceAndRouteOp(pos);  (3)

        if success then
          new_cost := ComputeCost(op);    (4)
          accepted := EvaluateNewPos();   (5)
          if accepted then
            break;
          else
            continue;
          endif                           (2)
      endfor

      if not accepted then
        RestoreOp();
      else
        CommitOp();

      if get a valid schedule then
        return scheduled;
    endfor

    if StopCriteria() then                (6)
      break;

    UpdateOverusePenalty();               (7)
    UpdateTemperature();                  (8)

  endwhile
  II++;
endwhile
```

Figure 2.6. Modulo scheduling algorithm for coarse-grained reconfigurable architecture

First all operations are ordered by the technique described in [11]. Priority is given to operations on the critical path and an operation is placed as close as possible to both its predecessors and successors, which effectively reduces the routing length between operations. Like other modulo scheduling algorithms, the outermost loop tries successively larger II, starting with an initial value equal to the minimal II (MII), until the loop has been scheduled. The MII is computed using the algorithm in [9].

For each II, our algorithm first generates an initial schedule which respects dependency constraints, but may overuse resources (1). For example, more than one operation may be scheduled on one FU in the same cycle. In the inner loop (2), the algorithm iteratively reduces resource overuse and tries to come up with a legal schedule. At every iteration, an operation is ripped up from the existing schedule, and is placed randomly (3). The connected nets are re-routed accordingly. Next, a cost function is computed to evaluate the new placement and routing (4). The cost is computed by accumulating the cost of all used MRRG nodes incurred by the new placement and routing of the operation. The cost function of each MRRG node is shown in eq. 2.2. It is constructed by taking into account the penalty of overused resources. In eq. 2.2, there is a basic cost (*base_cost*) associated with each MRRG node. The *occ* represents the occupancy of that node. The *cap* refers to the capacity of that node. Most MRRG nodes have a capacity of 1, whereas a few types of nodes such as the internal node of a register file have a capacity larger than one. The *penalty* factor associated with overused resources is increased at the end of each iteration (7). Through a higher and higher overuse penalty, the placer and router will try to find alternatives to avoid congestion. However, the penalty is increased gradually to avoid abrupt increases in the overused cost that may trap solutions into local minima. This idea is borrowed from the *Pathfinder* algorithm [8], which is used in FPGA P&R problems.

$$cost = base\_cost \times occ + (occ - cap) \times penalty \qquad (2.2)$$

In order to escape from local minima, we use a *simulated annealing* strategy to decide whether each move is accepted or not (5). In this strategy, if the new cost is lower than the old one, the new P&R of this operation will be accepted. On the other hand, even if the new cost is higher, there is still a chance that the move may be accepted, depending on the *"temperature"*. At the beginning, the temperature is very high so that almost every move is accepted. The temperature is decreased at the end of the each iteration (8). Therefore, the operation is increasingly difficult to move around. In the end, if the termination criteria is met without finding a valid schedule (6), the schedule algorithm starts with the next II.

2.4   Experimental Results

In the experiments, an architecture resembling the topology of MorphoSys [1] is instantiated from the ADRES template. In this configuration, a total of 64 FUs are divided into four tiles, each of which consists of 4 × 4 FUs. Each FU is not only connected to the 4 nearest neighbor FUs, but also to all FUs within the same row or column in this tile. In addition, there are row buses and column buses across the matrix. The first row of FUs is also used by the

Table 2.1. Scheduling results of kernels

Loop	No. of ops	Live-in vars	Live-out vars	II	IPC	Sched. density
idct1	93	4	0	3	31	48.4%
idct2	168	4	0	4	42	65.6%
adpcm-d	55	9	2	4	13.8	21.5%
mat_mul	20	12	0	1	20	31.3%
fir_cpl	23	9	0	1	23	35.9%

VLIW processor, and are connected to a multi-port register file. Only the FUs in the first row are capable of executing memory operations, i.e., load/store operations.

The testbench consists of 4 programs, which are derived from C reference code of TI's DSP benchmarks and MediaBench [14]. The *idct* is a 8 × 8 inverse discrete cosine transformation, which consists two loops. The *adpcm-d* refers to an ADPCM decoder. The *mat_mul* computes matrix multiplication. The *fir_cpl* is a complex FIR filter. They are typical multimedia and digital signal processing applications with abundant inherent parallelism.

The schedule results are shown in Table 2.1. The second column refers to the total number of operations within the loop body. The II is *initiation interval*. The live-in and live-out variables are allocated in the VLIW register file. The instructions-per-cycle (IPC) reflects how many operations are executed in one cycle on average. Scheduling density is equal to *IPC/No. of FUs*. It reflects the actual utilization of all FUs for computation. The results show the IPC is pretty high, ranging from 13.8 to 42. The FU utilization is ranged from 21.5% to 65.6%. For kernels such as *adpcm-d*, the results are constrained by achievable minimal II (MII). The Table 2.2 shows comparisons with the VLIW processor. The tested VLIW processor has the same configuration as the first row of the tested ADRES architecture. The compilation and simulation results for VLIW architectures are obtained from IMPACT, where aggressive optimizations are enabled. The results for the ADRES architecture are obtained from a developed co-simulator, which is capable of simulating the mixed VLIW and reconfigurable matrix

Table 2.2. Comparisons with VLIW architecture

App.	Total ops (ADRES)	Total cycles (ADRES)	Total ops (VLIW)	Total cycles (VLIW)	Speed-up
idct	211676	6097	181853	38794	6.4
adpcm_d	8150329	594676	5760116	1895055	3.2
mat_mul	20010518	1001308	13876972	2811011	2.8
fir_cpl	69126	3010	91774	18111	6.0

code. Although these testbenches are small applications, the results already reflect the impact of integrating the VLIW processor and the reconfigurable matrix. The speed-up over the VLIW is from 2.8 to 6.4, showing pretty good performance.

2.5 Conclusions and Future Work

Coarse-grained reconfigurable architectures have been gaining importance recently. Here we have proposed a new architecture called ADRES, where a VLIW processor and a reconfigurable matrix are tightly coupled in a single architecture. This level of integration brings a lot of benefits, including increased performance, simplified programming model, reduced communication cost and substantial resource sharing. We also describe a novel modulo scheduling algorithm, which is the key technology in the ADRES compiler. The scheduling algorithm is capable of mapping a loop to the reconfigurable matrix to expolit high parallelism. The experiment results show great performance advantage over the VLIW processor with comparable design efforts.

However, we have not implemented the ADRES architecture at the circuit level yet. Therefore, many detailed design problems have not been taken into account and concrete figures such as area and power are not available. Hence, to implement the ADRES design is in the scope of our future work. On the other hand, we believe the compiler is even more important than the architecture. We will keep developing the compiler to refine the ADRES architecture from the compiler point of view.

References

[1] H. Singh, et al. *MorphoSys: an integrated reconfigurable system for data-parallel and computation-intensive applications.* IEEE Trans. on Computers, 49(5):465–481, 2000

[2] C. Ebeling and D. Cronquist and P. Franklin. *RaPiD—Reconfigurable Pipelined Datapath.* Proc. of International Workshop on Field Programmable Logic and Applications (FPL), 1996

[3] PACT XPP Technologies. http://www.pactcorp.com

[4] T. Miyamori and K. Olukotun. *REMARC: Reconfigurable Multimedia Array Coprocessor.* International Symposium on Field Programmable Gate Arrays (FPGA), 1998

[5] B. Mei et al. *DRESC: A Retargetable Compiler for Coarse-Grained Reconfigurable Architectures.* International Conference on Field Programmable Technology, 2002

[6] D. A. Patterson and J. L. Hennessy. *Computer Architecture: A Quantitative Approach.* Morgan Kaufmann Publishers, Inc., 1996

[7] M. S. Lam. *Software pipelining: an effecive scheduling technique for VLIW machines.* Proc. ACM SIGPLAN '88 Conference on Programming Language Design and Implementation, 1988

[8] C. Ebeling et al. *Placement and Routing Tools for the Triptych FPGA.* IEEE Trans. on VLSI, 3(12):473–482, 1995

[9] B. Ramakrishna Rau. *Iterative Modulo Scheduling.* Hewlett-Packard Lab: HPL-94–115, 1995

[10] S. Roos. *Scheduling for ReMove and other partially connected architectures.* Laboratory of Computer Enginnering, Delft University of Technology, 2001

[11] J. Llosa et al. *Lifetime-Sensitive Modulo Scheduling in a Production Environment.* IEEE Trans. on Computers, 50(3):234–249, 2001

28

[12] C. Akturan and M. F. Jacome. *CALiBeR: A Software Pipelining Algorithm for Clustered Embedded VLIW Processors*. Proc. ICCAD, 2001

[13] M. M. Fernandes et al. *Distributed Modulo Scheduling*. Proc. High Performance Computer Architecture (HPCA), 1999

[14] C. Lee et al. *MediaBench: A Tool for Evaluating and Synthesizing Multimedia and Communicatons Systems*. International Symposium on Microarchitecture, 1997

Chapter 3

Stream-based XPP Architectures in Adaptive System-on-Chip Integration

Jürgen Becker[1], Martin Vorbach[2]

[1] *Universitaet Karlsruhe (TH)*
Institut fuer Technik der Informationsverarbeitung
D-76128 Karlsruhe, Germany
becker@itiv.uni-karlsruhe.de

[2] *PACT XPP Technologies AG*
Muthmannstr. 1
D-80939 Munich, Germany
martin.vorbach@pactcorp.com

Abstract   This chapter describes the structures and principles of stream-based, reconfigurable XPP Architectures from PACT XPP Technologies AG (Munich, Germany). The status of an adaptive System-on-Chip (SoC) integration is given, consisting of a SPARC-compatible LEON processor-core, a coarse-grain XPP-array of suitable size, and efficient multi-layer Amba-based communication interfaces. In addition PACTs newest XPP architectures are described, realizing a new runtime reconfigurable data processing technology that replaces the concept of instruction sequencing by configuration sequencing with high performance application areas envisioned from embedded signal processing to co-processing in different DSP-like and mobile application environments. The underlying programming model is motivated by the fact, that future oriented applications need to process streams of data decomposed into smaller sequences which are processed in parallel. Finally, first-time-right commercial chip synthesis were performed successfully onto 0.13 μm STMicro CMOS technology.

3.1   Introduction

Systems-on-Chip (SoCs) has become reality now, driven by fast development of CMOS VLSI technologies. Complex system integration onto one single die

P. Lysaght and W. Rosenstiel (eds.),
New Algorithms, Architectures and Applications for Reconfigurable Computing, 29–42.
© 2005 *Springer. Printed in the Netherlands.*

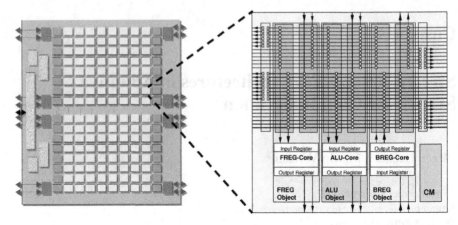

Figure 3.1. XPP64 Architecture Overview and Structure of one ALU PAE module

introduces a set of various challenges and perspectives for industrial and academic institutions. Important issues to be addressed here are cost-effective technologies, efficient and application-tailored hardware/software architectures, and corresponding IP-based EDA methods. Due to exponential increasing CMOS mask costs, essential aspects for the industry are now flexibility and adaptivity of SoCs. Thus, in addition to ASIC-based, one new promising type of SoC architecture template is recognized by several academic [2] [15] [16] [17] [18] [19] and first commercial versions [4] [5] [6] [7] [8] [9] [10] [12]: Configurable SoCs (CSoCs), consisting of processor-, memory-, probably ASIC-cores, and on-chip reconfigurable hardware parts for customization to a particular application. CSoCs combine the advantages of both: ASIC-based SoCs and multichip-board development using standard components [3]. This contribution provides the description of a CSoC project, integrating the dynamically reconfigurable eXtreme Processing Platform (XPP) from PACT [9] [10], [11] (see Figure 3.1). The XPP architecture realizes a new runtime re-configurable data processing technology that replaces the concept of instruction sequencing by configuration sequencing with high performance application areas envisioned from embedded signal processing to co-processing in different DSP-like application environments. The adaptive reconfigurable data processing architecture consists of following main components:

- Coarse-grain Processing Array Elements (PAEs), organized as Processing Arrays (PAs),

- a stream-based self synchronizing communication network,

- a hierarchical Configuration Manager (CM) tree, and

- a set of I/O modules.

This supports the execution of multiple data flow applications running in parallel. A PA together with one low level CM is referred as PAC (Processing Array Cluster). The low level CM is responsible for writing configuration data into the configurable objects of the PA. Typically, more than one PAC is used to build a complete XPP device. Doing so, additional CMs are introduced for configuration data handling. With an increasing number of PACs on a device, the configuration hardware assumes the structure of a tree of CMs. The root CM of the tree is called the supervising CM or SCM. This unit is usually connected to an external or global RAM. The basic concept consists of replacing the Von-Neumann instruction stream by automatic configuration sequencing and by processing data streams instead of single machine words, similar to [1]. Due to the XPPs high regularity, a high level compiler can extract instruction level parallelism and pipelining that is implicitly contained in algorithms [11]. The XPP can be used in several fields, e.g. as image/video processing, encryption, and baseband processing of next generation wireless standards, e.g. to realize also Software Radio approaches. 3G systems, i.e. based on the UMTS standard, will be defined to provide a transmission scheme which is highly flexible and adaptable to new services. Relative to GSM, UMTS and IS-95 will require intensive layer 1 related operations, which cannot be performed on today's processors [13] [14]. Thus, an optimized HW/SW partitioning of these computation-intensive tasks is necessary, whereas the flexibility to adapt to changing standards and different operation modes (different services, QoS, BER, etc.) has to be considered. Therefore, selected computation-intensive signal processing tasks have to be migrated from software to hardware implementation, e.g. to ASIC or coarse-grain reconfigurable hardware parts, like the XPP architecture.

3.2   Stream-based XPP Architecture

The XPP (eXtreme Processing Platform) architecture is based on a hierarchical array of coarse-grain, adaptive computing elements called Processing Array Elements (PAEs) and a packet-oriented communication network. The strength of the XPP technology originates from the combination of array processing with unique, powerful run-time reconfiguration mechanisms. Since configuration control is distributed over several Configuration Managers (CMs) embedded in the array, PAEs can be configured rapidly in parallel while neighboring PAEs are processing data. Entire applications can be configured and run independently on different parts of the array. Reconfiguration is triggered externally or even by special event signals originating within the array, enabling self-reconfiguring designs. By utilizing protocols implemented in hardware, data and event packets are used to process, generate, decompose and merge streams of data. The XPP has some similarities with other coarse-grain reconfigurable architectures like the KressArray [20] or Raw Machines [21] which are specifically for stream-based applications. XPP's main distinguishing features are its automatic packet-handling mechanisms and sophisticated hierarchical configuration protocols.

3.2.1 Array Concept and Datapath Structure

An XPP device contains one or several Processing Array Clusters (PACs), i.e. rectangular blocks of PAEs. Each PAC is attached to a CM responsible for writing configuration data into the configurable objects of the PAC. Multi-PAC devices contain additional CMs for configuration data handling, forming a hierarchical tree of CMs. The root CM is called the supervising CM or SCM. The XPP architecture is also designed for cascading multiple devices in a multi-chip. A CM consists of a state machine and internal RAM for configuration caching. The PAC itself contains a configuration bus which connects the CM with PAEs and other configurable objects. Horizontal busses carry data and events. They can be segmented by configurable switch-objects, and connected to PAEs and special I/O objects at the periphery of the device. A PAE is a collection of PAE objects. The typical PAE shown in Figure 3.1 contains a BREG object (back registers) and an FREG object (forward registers) which are used for vertical routing, as well as an ALU object which performs the actual computations. The ALU implemented performs common fixed-point arithmetical and logical operations as well as several special three-input opcodes like multiply-add, sort, and counters. Events generated by ALU objects depend on ALU results or exceptions, very similar to the state flags of a classical microprocessor. A counter, e.g., generates a special event only after it has terminated. The next section explains how these events are used. Another PAE object implemented in the prototype is a memory object which can be used in FIFO mode or as RAM for lookup tables, intermediate results and so on.

3.2.2 Stream Processing and Selfsynchronization

PAE objects as defined above communicate via a stream-based and selfsynchronizing bus and interface network. Two types of packets are sent through the array: data packets and event packets. Data packets have a uniform bit width specific to the device type. In normal operation mode, PAE objects are selfsynchronizing. An operation is performed as soon as all necessary data input packets are available. The results are forwarded as soon as they are available, provided the previous results have been consumed. Thus it is possible to map a signal-flow graph directly to ALU objects, and to pipeline input data streams through it. The communication system is designed to transmit one packet per cycle. Hardware protocols ensure that no packets are lost, even in the case of pipeline stalls or during the configuration process. Such consistent intra-array data and event transmission is one of the major differences between fine-grain CLB-based devices and coarse-grain XPP architectures, because application development is simplified considerably by mapping signal flow graphs directly onto the array topology. No explicit scheduling of operations and wire-based electrical signal as well as timing realization is required. Event packets are one bit wide. They transmit state information which controls ALU execution and packet generation. For instance, they can be used to control the merging of data-streams or to deliberately discard data packets. Thus conditional computations depending on the results of earlier ALU operations are feasible. Events can even trigger a self-reconfiguration of the device as explained below.

3.2.3 Configuration Handling

The XPP architecture is optimized for rapid and user transparent configuration. For this purpose, the configuration managers in the CM tree operate independently, and therefore are able to configure their respective parts of the array in parallel. Every PAE stores locally its current configuration state, i.e. if it is part of a configuration or not (states 'configured' or 'free'). If a configuration is requested by the supervising CM, the configuration data traverses the hierarchical CM tree to the leaf CMs which load the configurations onto the array. The leaf CM locally synchronizes with the PAEs in the PAC it configures. Once a PAE is configured, it changes its state to 'configured'. This prevents the respective CM from reconfiguring a PAE which is still used by another application. The CM caches the configuration data in its internal RAM until the required PAEs become available. Hence the CMs' cache memory and the distributed configuration state in the array enables the leaf CMs to configure their respective PACs independently. No global synchronization is necessary. While loading a configuration, all PAEs start to compute their part of the application as soon as they are in state 'configuredg. Partially configured applications are able to process data without loss of packets. This concurrency of configuration and computation hides configuration latency. Additionally, a prefetching mechanism is used. After a configuration is loaded onto the array, the next configuration may be requested and cached in the low-level CMs' internal RAM. For more details about the XPP architecture principles see [9], [10], [11].

3.3 Adaptive XPP-based System-on-Chip

The synthesized adaptive System-on-Chip is illustrated in Figure 3.2 and consists of an XPP-core from PACT, one LEON RISC μcontroller, and several

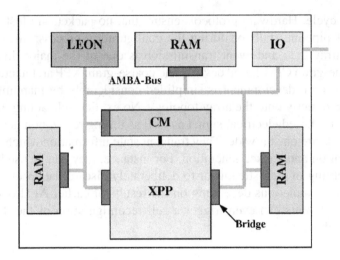

Figure 3.2. XPP-/Leon-based CSoC

SRAM-type memory modules. The main communication bus is chosen to be the AHB from ARM [22]. The size of the XPP architecture will be either 16 ALU-PAEs (4×4-array), or 64 ALU-PAEs (8x8-array), dependent on the application field (see section 3.4). To get an efficient coupling of the XPP architecture to AHB, we design an AHB-bridge which connects both IO-interfaces on one side of the XPP, input and output interfaces, to the AHB via one module. The AHB specification grants only communication between masters and slaves. There is no option provided for communication between two homogeneous partners, e.g. master to master. Usually the main controller, a processor or a μcontroller, on an AHB-based SoC is master, the RAM as a passive component on the bus is designed as a slave. Thus, if we choose our XPP-AHB-bridge to be a slave, there is no possibility for a connection between XPP and a RAM-module. Otherwise if we choose our bridge to be a master, no communication between the main μcontroller and XPP was allowed. Therefore we choose an unusual method and combine two ports, one master and one slave port as a dual port in the same bridge. This combination allows us to be flexible enough to process various application scenarios. In this way the XPP is able to handle the data from a RAM-module or gets a stream from another master on the CSoC. The CM unit implements a separate memory for faster storing and loading the XPP configurations. If there isn't enough memory space for storing the configurations in the local memory, it is possible to use the global CSoC memory to do that. The AHB-bridge for CM will be a single ported SLAVE-AHB-bridge. The transfers of the configurations from the global memory to the Configuration Manager will be done by the LEON. Therefore the CM has to send a request to LEON and start new configuration transfer. An overview on the CSoC architecture

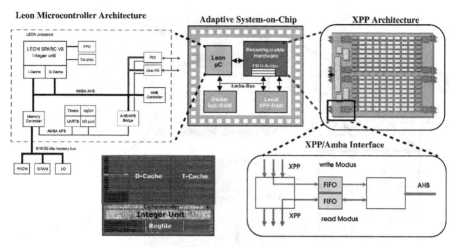

Figure 3.3. Adaptive XPP-based System-on-Chip Integration.

exemplarically integrated is shown in Figure 3.3. The μcontroller on our CSoC is a LEON processor [26]. This processor is a public domain IP core. The LEON VHDL model implements a 32-bit processor conforming to the SPARC V8 architecture synthesis. It is designed for embedded applications with the following features on-chip: separate instruction and data caches, hardware multiplier and divider, interrupt controller, two 24-bit timers, two UARTs, power-down function, watchdog, 16-bit I/O port and a flexible memory controller. Additional modules can be easily added using the on-chip AMBA AHB/APB buses. The VHDL model is fully synthesizable with most synthesis tools and can be implemented on both FPGAs and ASICs. The LEON μprocessor acts as a master on our CSoC architecture. The program data for LEON will be transferred via AHB. In this manner there are two options where the main memory for LEON could be located: internally on die or externally on separate modules. The local memory module on the integrated CSoC is used to store the LEON programs, data for XPP computation and XPP configurations. The theoretical bandwidth of AHB at 100 Mhz and 32bit bit width is 400 MBytes/sec. That's enough to serve the XPP-architecture with data and configurations and to handle the program data of the LEON efficiently. The interface of the memory module to the AHB is realized as a slave. That's because this module is a passive module only and can not start any kind of transactions on the AHB. Moreover, there will be an external RAM interface implemented, which allows to connect external memory to the CSoC. This module is a part of the LEON IP-core. The prior AHB specification [22] from ARM allows only one transaction per cycle. That means that if one master and one slave are communicating at a time step, the other modules on the bus have to wait till this communication is done. This kind of transactions block completely the whole bus. The solution for this restriction

36

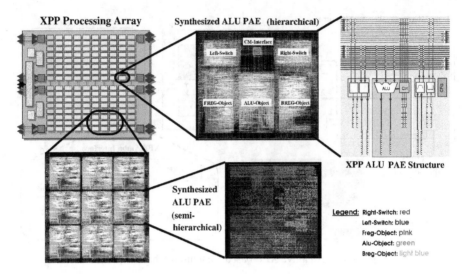

Figure 3.4. XPP ALU Structure and exemplaric Standard Cell Synthesis Layout at Universitaet Karlsruhe (TH).

is the multi-layer AHB. This concept allows multiple transactions at the same time. Instead of using simple multiplexers and decoders on the bus, we use now single decoders and multiplexers per slave to choose the right master for communication. Thus, the bus is divided in several sub-busses allowing simultaneous transactions between various masters and slaves. Each of the 4 parallel operating high-throughput bridges connecting the SRAM-banks to the XPP can achieve a data throughput of 400 MB/sec operating in 100 MHz, e.g. a complete on-Chip data throughput of 1600 MB/sec, which is sufficient for multimedia-based applications like MPEG-4 algorithms applied to video data in PAL-standard format (see section 3.5). The three SRAM memory modules provide up to 3 MB on-chip. The complete size is splitted into 2x1,2 MB and 1x0,6 MB modules, for storing two pictures and code with XPP configurations. Within this flexible multi-layer AHB interface concept the XPP can operate either as slave (Leon processor is master) or as master itself. The communication between the CSoC and the outside world will be realized through a master/slave AHB/PCI host bridge. The AHB master ability admits the direct transfers from PCI to internal RAM without involvement of the main μcontroller, or, the slave ability admits transfers between all masters and PCI-Bridge. Within this industrial/academic research project different exemplaric standard cell synthesis experiments and post-layout analysis were performed onto 0.18 and 0.13 μm UMC CMOS technologies at Universitaet Karlsruhe (TH) [23], [24]. The goal was to evaluate the trade-offs between CMOS design times for a modular or flattened synthesis of various application-tailored XPP-versions. Promising trade-offs between modularity of XPP hardware components and overall performance/power results have been obtained by applying so-called semi-hierarchical standard cell synthesis techniques, whereas only in the final backend place and route steps the hierarchy and borders of different hardware components were neglected (see Figure 3.4).

3.4   XPP64A: First-Time-Right-Silicon

The PACT XPP64A is a demonstrator implementation of PACT's XPP synthesizable cores. XPP cores are high performance reconfigurable processors for streaming applications in high performance reconfigurable accelerator/co-processor architecture solutions. Built as a demonstrator and a proof of concept, the XPP64A is, nevertheless, a complete fully functional product targeted at applications requiring extreme computational power. Produced in fully static 0.13 μm STMicro CMOS technology the XPP64A operates in speeds up to 64 MHz using standard 3.3V TTL logic compatible I/O pins (see Figure 3.5). Four independent I/O interfaces can each be configured as a RAM interface or as a pair of independent bidirectional streaming data channels using a simple hand-shaking protocol. ALU PAEs are organized in an 8x8 matrix and provide the majority of the computational capability. The ALU PAE provides a 24-bit DSP-compliant instruction set, whereas a (12,12)-bit opcode split is

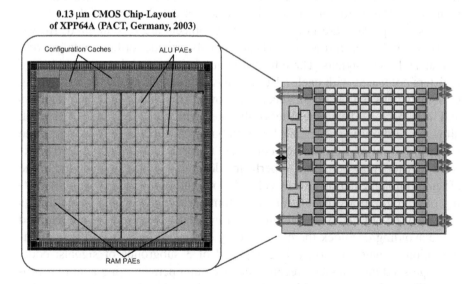

Figure 3.5. Coarse-grain Configurable XPP64A from PACT: STMicro CMOS 0.13 μm layout

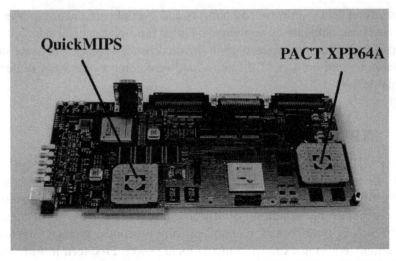

Figure 3.6. Coarse-grain Configurable XPP64A from PACT: STMicro CMOS 0.13 μm layout

available for performing complex addition/multiplication as well as conditional sign-flips. Single bit event signals allow status and control information such as carry to be processed. These event signals can also control other ALUs or request reconfiguration. Two rows of 8 RAM blocks are placed one each on the left and right of the processing array. Each RAM block is dual ported and organized as 256 words of 24 bits. The packet oriented nature of the network ensures that packets are never lost even as objects are reconfigured. Busses above and below each row of ALUs and RAMs provide for horizontal routing of data and event signals. Three forward data registers and three forward event registers colocated with each ALU and RAM block allow downward routing of signals between busses in successive rows. Three backward data registers and three backward event registers collocated with each ALU and RAM block allow upward routing of signals between busses in successive rows. The three backward event registers can be combined to form a lookup table. Two backward registers may be paired to perform addition or subtraction and aid in the efficient implementation of FIR cells. A configuration manager handles writing of configurations into the array. Automatic resource management allows unconfigured objects to be configured before all needed resources are required while avoiding deadlock thereby speeding configurations. The XPP64A configuration manager interface is comprised of 3 subgroups of signals: code, message send and message receive. Each group operates using a similar protocol. The code group provides a channel over which code can be downloaded to the XPP64A for caching or execution. The message send group provides a channel over which the XPP64A can send messages to a host microprocessor and the message receive group provides a channel for a host microprocessor to send messages to the XPP64A. The XPP64A represents a commercial first-time-right-silicon solution onto $0.13\mu m$ STMicro CMOS technology, whereas backend synthesis and place&route steps were successfully supported by the STMicroelectrics daughter company ACCENT (Milano, Italy). In addition, an XXP evaluation board has been developed and fabricated in parallel in order to evaluate and apply the XPP64A system (see Figure 3.6). For more details see [9].

3.5 Application Evaluation—Examples

Within the application area of future mobile phones desired and important functionalities are gaming, video compression for multimedia messaging, polyphone sound (MIDI), etc. Therefore, a flexible, low cost hardware platform with low power consumption is needed for realizing necessary computation-intensive algorithms parts. Actual applications currently mapped and analyzed onto an 4x4 ALU PAE array (extracted from the XPP64A) are stream-based and computational intensive algorithm parts from MPEG-4 algorithms, W-LAN (Hiperlan-2), Software Radio including RAKE-Receiver and Baseband Filtering. The obtained performance and power efficiency improvements are compared to a state of the art digital signal processor [25]. Thus, PACT implemented several functionalities onto the cost-efficient 4x4 XPP array size, e.g. a 256-point FFT, a real 16 tap FIR filter, and a video 2D-DCT (8x8) for MPEG-4 systems (see Figure 3.7). The last functionality will be discussed here. The performance and power consumption benchmarks comparisons between XPP and a state-of-the-art DSP Processor for the above mentioned functionalities are given in Figure 3.8. First digital TV application performance results were obtained by evaluating corresponding MPEG-4 algorithm mappings onto the introduced XPP64A and based on the 0.13 μm CMOS technology synthesis results. The Inverse DCT was applied to 8x8 pixel blocks and have been performed by an 4x4 XPP-Array being sufficient for real-time digital

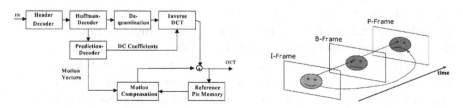

Figure 3.7. Main MPEG-4 Algorithm Overview and main Modules.

Algorithm	Performance	Power Consumption
2D DCT (8x8) 10,000 Blocks	640,000 Cycles **25 ms @ 26 MHz**	0.4 mA / MHz **9 mW @ 0.9V; 26 MHz**
1D DCT (32) 1024 Points	2048 Cycles **79 μs @ 26 MHz**	0.2 mA / MHz **5 mW @ 0.9V; 26 MHz**
Viterbi K = 9 100 Samples	12800 Cycles **492 μs @ 26 MHz**	0.2 mA / MHz **5 mW @ 0.9V; 26 MHz**
FFT 64 Points (packed 12,12 bits)	280 Cycles **5 μs @ 52 MHz**	0.5 mA / MHz **13.7 mW @ 0.9V; 52 MHz**
FIR 16 Taps 1000 Samples	1000 Cycles **38 μs @ 26 MHz**	0.5 mA / MHz **12 mW @ 0.9V; 26 MHz**

4x4 ALU PAE array with 2x4 RAM PAEs

Figure 3.8. Application Evaluation Results based on the XPP64A Silicon.

TV application scenarios, from the performance as well as from the power consumption requirements.

3.6 Conclusions

This chapter described the stream-based and selfsynchronizing XPP architecture principles as well as corresponding coarse-grain hardware structures. The focus was given to its target platform and Configurable System-on-Chip (CSoC) integration concepts and solutions. An academic/industrial CSoC project was discussed by exemplaric synthesis experiments performed at Universitaet Karlsruhe (TH). Here, an overview on the overall system concept, the hardware datapath structures and their integration was provided. Next, PACTs latest commercial XPP chip and platform variants have been introduced, synthesized successfully onto 0.13 μm STMicro CMOS technologies. Here, the XPP64A high-performance reconfigurable accelerator solution was presented, an adaptive 8x8 ALU Array operating as co-processor for applications with tremendous performance requirements, e.g. mobile communication basestations, computing fabrics, etc. The XPP64A represents first-time-right-silicon demonstrating and proofing the feasibility and possibilities of PACTs leading edge coarse-grain reconfigurable technology. Finally, some evaluated application performance results within the area of multimedia and mobile baseband algorithms have been discussed, e.g. the implementation of a video 2D-DCT (8x8) in MPEG-4 systems suitable for mobile multimedia messaging. In summary, PACT offers the first commercially relevant silicon and computing platforms within the promising field of coarse-grain reconfigurable computing. This technology provides similar flexibilities as DSPs or other application processors, with much better performance and power (energy) consumption trade-offs. Compared to ASICs, XPP-based System-on-Chip solutions are less risky and much cheaper from the development and mask costs. In addition, a great potential for multi-service and multi-standard application is provided resulting in multi-purpose adaptivity and higher volumes.

References

[1] R. W. Hartenstein, J. Becker et al.: *A Novel Machine Paradigm to Accelerate Scientific Computing*; Special issue on Scientific Computing of Computer Science and Informatics Journal, Computer Society of India, 1996.

[2] J. Becker, T. Pionteck, C. Habermann, M. Glesner: *Design and Implementation of a Coarse-Grained Dynamically Reconfigurable Hardware Architecture*; in: Proc. of IEEE Computer Society Annual Workshop on VLSI (WVLSI 2001), Orlando, Florida, USA, April 19-20, 2001.

[3] J. Becker (Invited Tutorial): *Configurable Systems-on-Chip (CSoC)*; in: Proc. of 9th Proc. of XV Brazilian Symposium on Integrated Circuit Design (SBCCI 2002), Porto Alegre, Brazil, September 5-9, 2002.

[4] Xilinx Corp.: http://www.xilinx.com/products/virtex.htm.

[5] Altera Corp.: http://www.altera.com

[6] Triscend Inc.: http://www.triscend.com

[7] Triscend A7 Configurable System-on-Chip Platform—Data Sheet http://www.triscend.com/products/dsa7csoc_summary.pdf

[8] LucentWeb http://www.lucent.com/micro/fpga/

[9] PACT Corporation: http://www.pactcorp.com

[10] The XPP Communication System, PACT Corporation, Technical Report 15, 2000.

[11] V. Baumgarte, F. Mayr, A. Nückel, M. Vorbach, M. Weinhardt: *PACT XPP—A Self-Reconfigurable Data Processing Architecture*; The 1st Intl. Conference of Engineering of Reconfigurable Systems and Algorithms (ERSA01), Las Vegas, NV, June 2001.

[12] Hitachi Semiconductor: http://semiconductor.hitachi.com/news/triscend.html

[13] Peter Jung, Joerg Plechinger.:*M-GOLD: a multimode basband platform for future mobile terminals*, CTMC'99, IEEE International Conference on Communications, Vancouver, June 1999.

[14] Jan M. Rabaey: *System Design at Universities: Experiences and Challenges*; IEEE Computer Society International Conference on Microelectronic Systems Education (MSE99), July 19–21, Arlington VA, USA.

[15] S. Copen Goldstein, H. Schmit, M. Moe, M. Budiu, S. Cadambi, R. R. Taylor, R. Laufer: *PipeRench: a Coprocessor for Streaming Multimedia Acceleration* in ISCA 1999. http://www.ece.cmu.edu/research/piperench/

[16] MIT Reinventing Computing: http://www.ai.mit.edu/projects/transit dpga_prototype_documents.html

[17] N. Bagherzadeh, F. J. Kurdahi, H. Singh, G. Lu, M. Lee: *Design and Implementation of the MorphoSys Reconfigurable Computing Processor*; J. of VLSI and Signal Processing-Systems for Signal, Image and Video Technology, 3/2000.

[18] Hui Zhang, Vandana Prabhu, Varghese George, Marlene Wan, Martin Benes, Arthur Abnous, *A 1V Heterogeneous Reconfigurable Processor IC for Baseband Wireless Applications*, Proc. of ISSCC2000.

[19] Pleiades Group: http://bwrc.eecs.berkeley.edu/Research/Configurable _Architectures/

[20] R. Hartenstein, R. Kress, and H. Reinig *A new FPGA architecture for word-oriented datapaths* In Proc. FPL'94, Prague, Czech Republic, September 1994. Springer LNCS 849.

[21] E. Waingold, M. Taylor, D. Srikrishna, V. Sarkar, W. Lee, V. Lee, J. Kim, M. Frank, and P. Finch: *Baring it all to software: Raw machines* IEEE Computer, pages 86-93, September 1997.

[22] ARM Corp.: http://www.arm.com/arm/AMBA

[23] J. Becker, M. Vorbach: *Architecture, Memory and Interface Technology Integration of an Industrial/Academic Configurable System-on-Chip (CSoC)*; IEEE Computer Society Annual Workshop on VLSI (WVLSI 2003), Tampa, Florida, USA, February, 2003.

42

[24] J. Becker, A. Thomas, M. Vorbach, V. Baumgarte: *An Industrial/Academic Configurable System-on-Chip Project (CSoC): Coarse-grain XPP-/Leon-based Architecture Integration*; Design, Automation and Test in Europe Conference (DATE 2003), Munich, March 2003.

[25] M. Vorbach, J. Becker: *Reconfigurable Processor Architectures for Mobile Phones*; Reconfigurable Architectures Workshop (RAW 2003), Nice, France, April, 2003.

[26] Jiri Gaisler: *The LEON Processor User's Manual*, http://www.gaisler.com

Chapter 4

Core-Based Architecture for Data Transfer Control in SoC Design

Unai Bidarte, Armando Astarloa, Aitzol Zuloaga, José Luis Martín and Jaime Jiménez

University of the Basque Country, E.T.S. Ingenieros, Bilbao—SPAIN
{jtpbipeu, jtpascua, jtpzuiza, jtpmagoj, jtpjivej}@bi.ehu.es

Abstract   The functionality of many digital circuits is strongly dependent on high speed data exchange between data source and sink elements. In order to alleviate the main processor's work, it is usually interesting to isolate high speed data transfers from all other control tasks. A generic architecture, based on configurable cores, has been achieved for circuits communicating with an external system and with extensive data exchange requirements. Design reuse has been improved by means of a software application that helps in architectural verification and performance analysis tasks. Two applications implemented on FPGA technology are presented to validate the proposed architecture.

4.1   Introduction

When analyzing the data path of a generic digital system, four main elements can be distinguished:

1. Data source, processor or sink. Any digital system needs some data source, which can be some kind of sensor or an external system. Process units transform data and finally data clients or sinks make some use of it.

2. Data buffer blocks or memories. The data units cannot usually be directly connected and intermediate storage elements are needed.

3. Data transfer control unit. On the other hand, when transferring data from one device to another, a communication channel and a predefined data transfer protocol are needed. Data transfer control can be a very time

P. Lysaght and W. Rosenstiel (eds.),
New Algorithms, Architectures and Applications for Reconfigurable Computing, 43–54.
© 2005 *Springer. Printed in the Netherlands.*

consuming task. In order to liberate the main control unit, it is often necessary to to use a dedicated unit to control data transfer.

4. High level control unit. It is the digital system master that generates all the control signals needed by the previously noted blocks.

This work studies digital systems with much data exchange, that is to say, circuits with high volume data transfers. The following features summarize the system under study:

- There is a communication channel with an external system, that is used to transmit and receive control commands and data. Control commands are exchanged between main processors (high level control units, one per system), always through the communication channel. Data goes from a data source to a data sink and the communication channel is needed whenever the source and the sink elements are not in the same system.

- Data processing is performed on data terminal units, so, for design purposes, data units are supposed to be sources or sinks. These elements are not synchronized, so intermediate data storage is needed.

- It is needed to attend several data transfer requirements in parallel.

- Data exchange requires complex and high speed control, which makes a specific data transfer module necessary. The bus used to perform high speed data transfers and the bus used by the main processor to perform high level control tasks work independently but there is a bridge to connect them, because the main processor receives control commands through the communication channel.

Figure 4.1 represents the block diagram corresponding to the described circuit. Several applications match the specifications above: industrial machinery like filling or milling machines, polyphonic audio, generation of three dimensional images, video servers, and PC equipment like plotters and printers.

The research team's main objective is to achieve a reusable architecture for such systems. It must be independent of the implementation technology, so a standard hardware description language will be used. In order to facilitate design reuse, a user friendly software application will be programmed to make the verification and performance analysis easier.

4.2 Digital Systems with Very Time Consuming Data Exchange Requirements. Design Alternatives

Traditionally, embedded systems like the one under study have been successfully developed with complex 16 or 32 bits microcontrollers. These process

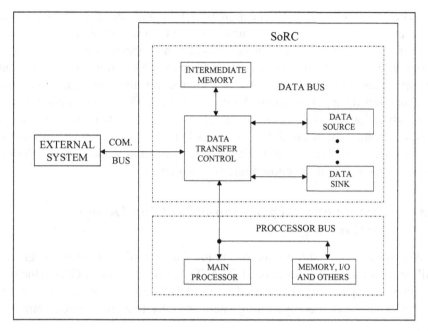

Figure 4.1. Block diagram of the generic circuit.

machines perform millions of instructions per second, and include some communication channels, memory interfaces, and direct memory access controllers. On the other hand, they present many disadvantages that make impossible to fulfill our design goals:

- As they are general purpose integrated circuits, no feature can be adapted to the application. Sometimes software patches will substitute for hardware requirements.

- Although they have different communication interfaces, no frame protocol encoding or decoding is usually available, so these tasks become software work.

- No data source or sink interfaces are available, so software and general purpose input and output ports are used.

- These features force the high level control machine to perform data exchange control tasks, so low speed work is seriously limited. An external circuit dedicated to these tasks can be used to alleviate the data exchange control bottleneck.

An optimum solution requires an architecture focused on high speed data exchange performed in an asynchronous mode between source and sink elements

[1]. The complete system has been achieved on one chip [2]. The design is modular and based on parameterizable cores to facilitate future reuse [3].

There is an intermediate solution between the general purpose microcontroller based solution and the System-on-Chip (SoC) solutions. High level and low speed tasks can be performed by a microcontroller and high speed data exchange left for an autonomous hardware system. This solution can be adequate when features that are not available on FPGAs but that are common on microcontrollers are needed. Examples include A/D or D/A data converters, FLASH and EEPROM storage. This is a less integrated and slower solution and it is dependent on the chosen microcontroller.

4.3   System on a Reprogrammable Chip Design Methodology

With today's deep sub-micron technology, it is possible to deliver over two million usable system gates in one FPGA. The availability of FPGAs in the one million system gate range has started a shift of SoC designs towards using reprogrammable FPGAs, thereby starting a new era of System on a Reprogrammable Chip (SoRC) [4].

Today's market expects better and cheaper designs. The only way the electronics industry can achieve these needs in a reasonable amount of time is with design reuse. Reusable modules are essential to design complex circuits [5].

So the goal of this work is to achieve a modular, configurable and reusable architecture that performs very high speed data exchange without compromising the tasks performed by the main processor. Hardware and software co-design and co-verification are also important objectives [6].

Traditionally IP cores used non-standard interconnection schemes that made them difficult to integrate. This required the creation of custom glue logic to connect each of the cores together. By adopting a standard interconnection scheme, the cores can be integrated more quickly and easily by the end user. A standard data exchange protocol is needed in order to facilitate SoRC design and reuse. Excluding external system buses such as PCI, VME, and USB, there are many SoC interconnection buses. Most of them are proprietary. Examples include the Advanced Microcontroller Bus Architecture (AMBA, from ARM) [7], CoreConnect (from IBM) [8], FISPbus (from Mentor Graphics and Inventra Business Unit) [9], and the IP interface (from Motorola). [10]. We looked for an open option, that is to say, a bus which does not need any license agreement and with no need to pay any kind of royalty.

The solution is the Wishbone SoC interconnection architecture for portable IP cores. The Wishbone standard defines the data exchange among IP core modules, and it does not regulate the application specific functions of the IP core [11]. It offers a flexible integration solution, a variety of bus cycles and data path widths to solve various system problems, and allows cores to be designed by a variety of designers. It is based on a master/slave architecture for very flexible system designs. All Wishbone cycles use a handshaking protocol between Master and Slave interfaces.

4.4   SoRC Core-Based Architecture

The SoRC architecture shown in Figure 4.2 complies with the specifications noted. The main objective of the architecture is to isolate high speed data exchange from any other control tasks in the system. That is why the design has been divided into three blocks, each one with its own specific bus:

- The "Data Transfer Bus" (DTB) is used to perform all process data transfers and it uses the Wishbone SoC interconnection architecture for portable IP cores, which makes possible high speed data exchange. A shared bus interconnection with only one master has been chosen, which controls all transfers in the bus.

- The "Main Processor Bus" (MPB) is used to perform high level tasks, which are controlled by a high level controller or main processor, usually a

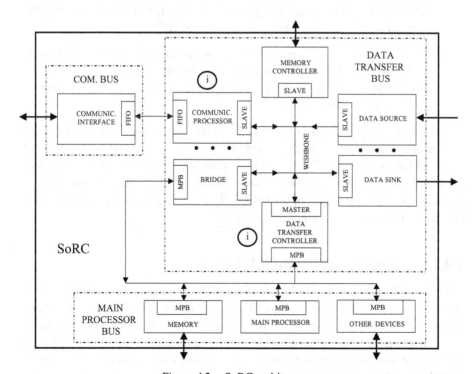

Figure 4.2.   SoRC architecture.

microcontroller. It uses its specific bus to read from and write to program memory, input and output blocks and any other devices, as well as to communicate with the DTB.

- The "Communication Bus" (CB) is used to exchange information with the external system. It cannot directly access the DTB and it exchanges information with the communication processor, which is a module on the DTB.

On the other hand, the SoRC has these connections with the outside: the communication channel interface, the memory bus, the data source and/or sink devices interface and the interface to devices on the main processor side.

4.4.1 Communication Bus IP Cores

The communication interface must provide the interface to the communication channel. After dealing with a number of different communication channels, the solution we have chosen is:

- If the communication interface needs a very complex controller (Bluetooth, USB, etc.) and there is an adequate solution available in ASSP or ASIC format, it is useful and practical to use it. In those cases only the channel interface is implemented into the SoRC.

- For non-complex communication protocols (UARTs, parallel port buses such as IDE or EPP.) both the controller and the interface are embedded into the SoC architecture leaving only the physical drivers and protection circuits outside the FPGA.

All the developed cores support full duplex communication. These cores have a common interface to the communication processor, which consists of two FIFO memories, one for reception and the other one for transmission. The communication interface presents two control signals to the communication processor to show the status of the FIFOs. The communication processor reads or writes the memories whenever it is needed.

4.4.2 Data Transfer Bus IP Cores

DATA TRANSFER CONTROLLER (DTC): This core allows data transfers between any two Wishbone compatible modules. It is the unique Wishbone master module, so it controls all operations on the bus. The system critical task is high speed data exchange, which must be performed in parallel between different origin and destination pairs. To complete any transfer, the DTC must read the data from the origin and then write it to the destination. Many transfer requests can be activated concurrently, so the DTC must be capable of serving them. In order to guarantee that no request is blocked by another one, any pending data transfer request attention is initiated only when the origin and destination cores are ready. The DTC priories the requests following a round robin scheme.

The key to the control is to manage the right number of data channels, which must be exactly the number of concurrent data movements that can be accepted. Figure 4.3 summarizes the solution adopted for the hypothetical case with one data source and one data sink. The minimum data channels are as follows:

1. From the communication processor to the main processor bridge, in case of commands reception (from the external system), or to the memory, if data for a sink is received.

2. From the main processor bridge to the communication processor, in case of commands transmission (to the external system), or from the memory to the communication processor, in case of data transmission.

3. From the memory to a data sink.

4. From a data source to the memory.

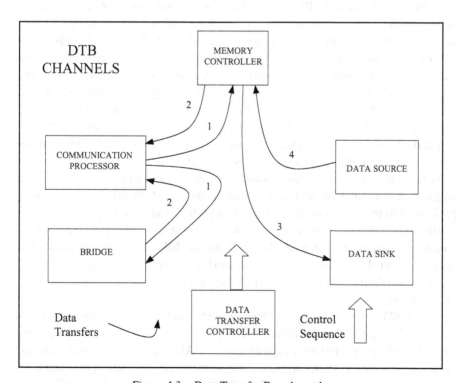

Figure 4.3. Data Transfer Bus channels.

The minimum number of channels is two plus one additional channel for each data source or sink. Each channel has three control registers: origin, destination and transfer length. Some of them are fixed and others must be configured by the main processor before starting data exchange. Additionally, there are two common registers which contain one bit associated with each channel: the control register to enable or disable transfers, and the start register, which must be asserted by the main processor, after correctly configuring the three registers of the channel, to start the data transfer.

An interrupt register is used to acknowledge the termination of the data exchange to the main processor. Once the requested data transfer is accomplished, the DTC asserts the interrupt register bit associated with the channel. There is only one interrupt line, so whenever an interrupt occurs, the main processor must read the interrupt register, process it and deactivate it.

Partial address decoding has been used, so each slave decodes only the range of addresses that it uses. This is accomplished using an address decoder element, which generates all chip select signals. The advantages of this method are that it facilitates high speed address decoding, uses less redundant address decoding logic, supports variable address sizing and supports variable interconnection schemes.

COMMUNICATION PROCESSOR: The frame information contains a header, commands for the main processor, file configuration, data and a check sequence. The receiver part of this module decodes data frames and sends data to the correct destination under the DTC control. The transmitter part is responsible for packaging the outgoing information. This core permits full duplex communication so the receiver and transmitter parts are completely independent.

The main processor must know about command reception because it configures all transfers. This core generates two interrupts, one when a command is received and another when communication error is detected. These are the two only interruptions not generated by the DTC.

When data intensive communication is performed, some kind of data correctness check must be performed. The communication processor is able to perform different kinds of check sequence coding and decoding.

BRIDGE: The main processor cannot access data on the DTB directly, so an intermediate bridge between the DTB and the MPB is needed. It must adapt data and control interfaces. Usually the data bus on the DTB is wider than on the MPB so one data transfer on the DTB corresponds to more than one operation on the MPB.

MEMORY CONTROLLER: Data exchange between data source and sink elements is supposed to be performed asynchronously. This is possible using an intermediate data buffer. Large blocks of RAM memory are needed in data exchange oriented systems and stand-alone memories provide good design solutions. The design must be capable of buffering several megabytes of data, so dynamic memory is needed and in order to optimize memory access it must also be synchronous. Consequently a synchronous and dynamic memory (SDRAM) controller has been developed [12].

High speed systems like the one presented here must follow synchronous methodology rules. The generation, synchronization and distribution of clock signals are essential. FPGAs designed for SoRC provide high speed, low skew clock distributions through dedicated global routing resources and Delay Locked Loop (DLL) circuits. A DLL works by inserting delay between the input clock and the feedback clock until the two rising edges align. It is not possible to use one DLL to provide both the FPGA and SDRAM clocks. Using two DLLs with the same clock input and separate feedback signals achieves zero delay between the input clock, the FPGA clock, and the SDRAM clock.

DATA SOURCE/SINK: Data from a source is written to a sink, using the intermediate memory and the communication channel when required. The data source and sink control is performed by application dependent cores.

4.4.3 Main Processor Bus IP Cores

This main processor and all the resources it requires are connected to the MPB. The MPB can also be implemented using the Wishbone standard. We have assumed that the circuit is oriented to data exchange, which means that this task is very time consuming and it justifies a specific data exchange control unit. All other tasks can be controlled by a general purpose machine, usually a microcontroller, and it will be chosen in accordance with the application. This multifunction machine uses its own bus to access memory, input/outputs, user interface, other devices, and the DTB.

A command reception interrupt from the communication processor tells the main processor about the request from the external system. The main processor reads it from the bridge, through the DTC, and processes it. If it is a control command, it will send back the answer command. If it is a data command, it will configure the corresponding data channel on the DTC and after this it will send back the acknowledge or answer command to the external system. The DTC core will generate the data transfer end interrupt when this operation is finished.

Whenever it is detected that data coming from the external system is corrupted, the communication processor activates the error interrupt. The main processor will tell back the external system about the failure.

Commands and data information exchange can also be initiated by this circuit, so the architecture allows information exchange between two systems in a symmetrical way.

4.5 Verification and Analysis User Interface

In order to improve the reusability and to make it easier to verify and analyze the proposed architecture, a user interface application has been developed [13].

The use of a hardware description language like VHDL has facilitated the creation of a parameterizable design. The specifications can be kept open and design alternatives can be evaluated, due to the fact that the design parameters can be modified. Design modularity and cores parameterization greatly improve future reuse possibilities.

To do a generic design, the effort needed at the beginning of the project is bigger than for a fixed design, in which component functionality is fully determined. But this technique, apart from the advantages mentioned above, could greatly alleviate the unexpected problems that arise in the final stages of the design process.

Some hard coded values in the design have been replaced with constants or generics. In this way, even if the parameter is not going to be changed in the future, code readability is increased. A global package containing the definition of all parameters has been used. The designer can configure the application specific architecture writing to the global package or using the software interface.

Once the desired architecture is configured, and after designing the application specific cores, the complete system functionality must be validated. All the cores, as well as the high level control machine code, must be co-simulated and co-verified. Modelsim from Mentor Graphics is the simulation tool used. It provides a Tool Command Language and Toolkit (Tcl/Tk) environment offering script programmability. A custom testbench has been created to facilitate the visualization of simulation results. A Tcl/Tk program creates a new display based on the simulator's output data, where a selection of the signals can be visualized with data extracted from the simulation results. Some buttons have been added so that new functionality is accessible. Dataflow can be graphically analyzed and design debugging is much easier. The external system, communication channel, SDRAM and data source and sink functionality have been described using VHDL behavioural architectures. Data transfers on Wishbone bus are automatically analyzed by a supervisor core, which dramatically simplifies simulation. The supervisor also generates text files with performance information, including data transfer bit rate and latency.

4.6   Results and Conclusions

This section describes the application of the proposed architecture to the design of two digital systems.

The first one is a video system connected to a host via an Ethernet communication channel. The system controls two motors used for camera movement and processes incoming control commands. The DTC core performs image data exchange between the analog to digital converter and the host. The SDRAM is used as a ping-pong memory: while a video frame is being captured, the previous one is being transmitted to the host. A general purpose evaluation

Table 4.1. Implementation results

Features	Video processor	Industrial plotter
Equivalent gates	127.000	182.000
Internal RAM bits	16 Kbits	28 Kbits
User I/O pins	85	94
Max. DTC freq.	47 MHz	65 MHz
Main Processor	Nios 32 bits	MicroBlaze 32 bits

board from Altera containing the 20K200EFC484 device has been used for prototyping. System features are summarized in Table 4.1.

The second one corresponds to an industrial plotter that provides high efficiency on continuously working environments. The microcontroller manages one stepping motor, three dc motors, many input/output signals and a simple user interface. The DTC module is dedicated to high volume data exchange from the host to the printer device using a parallel communication channel. Printing information is buffered in the SDRAM. This circuit is based on a Spartan II family device from Xilinx which offers densities up to 200.000 equivalent gates. System features are summarized in Table 4.1.

The size, speed, and board requirements of today's state-of-the-art FPGAs make it nearly impossible to debug designs using traditional logic analysis methods. Flip-chip and ball grid array packaging do not have exposed leads that can be physically probed. Embedded logic analysis cores have been used for system debugging [14].

The results show that our SoRC architecture is suitable for generating hardware/software designs for digital systems with much data exchange. The multicore architecture with configuration parameters is oriented to reuse and a user friendly software application has been developed to help with application specific configuration, system hardware/software co-verification and performance analysis. The use of a standard and open SoC interconnection architecture improves the portability and reliability of the system and results in faster time to market.

Acknowledgments

This work has been supported by the Ministerio de Ciencia y Tecnologa of Spain and FEDER in the framework of the TIC2001-0062 research project.

References

[1] Cesrio, W., and Baghdadi, A. (2002). Component-Based Design Approach for Multicore SoCs. *Design Automation Conference, 39th conference*. New Orleans, Louisiana.

[2] Bergamaschi, R., and Lee, W. (2000). Designing Systems on Chip Using Cores. *Design Automation conference, 37th Conference*.

[3] Gupta, R., and Zorian, Y. (1997). Introducing Core-Based System Design. *IEEE Design and Test of Computers*, 14(4): 15–25.

[4] Xilinx. *Design Reuse Methodology for ASIC and FPGA Designers.* http://www.xilinx.com/ipcenter/designreuse/docs/ Xilinx-Design-Reuse-Methodology.pdf.

[5] Keating, M., and Bricaud, P. (2002). *Reuse Methodology Manual.* Kluwer Academic Publishers.

[6] Balarin, F. (1997). *Hardware-Software Co-design of Embedded Systems: The POLIS Approach.* Kluwer Academic Publishers.

[7] ARM. (1999). *AMBA Specification Revision 2.0.* http://www.arm.com/ Pro+Peripherals/AMBA.

[8] *IBMCoreConnect Bus Architecture.* http://www-3.ibm.com/chips/techlib/techlib.nsf/productfamilies/CoreConnect- Bus-Architecture.

[9] Mentor Graphics. *FISPbus.* http://www.mentorg.com.

[10] Motorola. *IP Interface, Semiconductor Reuse Standard, v. 2.0.* http:// www.motorola.com.

[11] *Wishbone System-on-Chip (SoC) Interconnection Architecture for Portable IP Cores (rev. B.2).* (2001). Silicore Corporation, Corcoran (USA).

[12] Gleerup, T. (2000). Memory Architecture for Efficient Utilization of SDRAM: A Case Study of the Computation/Memory Access Trade Off. *Intl Workshop on Hardware-software Codesign*, 51–55.

[13] Bergeron, J. (2001). *Writing Testbenches. Functional Verification of HDL Models.* Kluwer Academic Publishers.

[14] Zorian, Y., Marinissen, E. J., and Dey, S. (2001). Optimal Test Access Architectures for System-on-a-Chip. *Transactions on Design Automation of Electronic Systems*, 26–49.

Chapter 5

Customizable and Reduced Hardware Motion Estimation Processors

Nuno Roma, Tiago Dias, Leonel Sousa

Instituto Superior Técnico/INESC-ID
Rua Alves Redol, 9—1000-029 Lisboa—PORTUGAL
Nuno.Roma@inesc-id.pt, tdias@sips.inesc-id.pt, las@inesc-id.pt

Abstract   New customizable and reduced hardware core-based architectures for motion estimation are proposed. These new cores are derived from an efficient and fully parameterizable 2-D systolic array structure for full-search block-matching motion estimation and inherit its configurability properties in what concerns the macroblock and search area dimensions and parallelism level. A significant reduction of the hardware resources can be achieved with the proposed architectures by reducing the spatial and pixel resolutions, rather than by restricting the set of considered candidate motion vectors. Low-cost and low-power regular architectures suitable for field programmable logic implementation are obtained without compromising the quality of the coded video sequences. Experimental results show that despite the significant complexity level presented by motion estimation processors, it is still possible to implement fast and low-cost versions of the original architecture using general purpose FPGA devices.

Keywords: Motion Estimation, Customizable Architectures, Configurable Structures

5.1   Introduction

Motion estimation is a fundamental operation in motion-compensated video coding [1], in order to efficiently exploit the temporal redundancy between consecutive frames. Among the several possible approaches, block-matching is the most used in practice: the current frame is divided into equally sized $N \times N$ pixel blocks that are displaced within $(N + 2p - 1) \times (N + 2p - 1)$ search windows defined in the previous frame; the motion vectors are obtained by looking for the best matched blocks in these search windows. In this procedure,

55

the Sum of the Absolute Differences (SAD) is the matching criteria that is usually used by most systems, due to its efficiency and simplicity.

Among the several block-matching algorithms, the Full-Search Block-Matching (FSBM) method is the one that has been used by most VLSI motion estimation architectures that have been proposed over the last few years. The main reason for this is not only related to the better performance levels generally achieved by exhaustively considering all possible candidate blocks, but is mainly due to the regularity properties that it also offers. In fact, not only does it lead to much more efficient hardware structures, but it also produces significantly simpler control units, which is always a fundamental factor towards a real-time operation based on hardware structures. However, it requires a lot of computational resources. As an example, FSBM motion estimation can consume up to 80% of the total computational power required by a video encoder. This fact often prevents its implementation using low cost technologies with restricted amounts of hardware (such as Field Programmable Logic (FPL) devices) and usually demands the usage of technologies with higher densities of gates, such as ASIC or Sea-of-Gates. As a consequence, several fast block-matching motion estimation algorithms have been proposed over the last years. Most of them restrict the search space to a given search pattern, providing suboptimal solutions (e.g. [2, 3]). Nevertheless, they usually apply non-regular processing and require complex control schemes, making their hardware implementation difficult and rather inefficient.

Hence, a tradeoff is frequently explored between the required computational resources, the implementation complexity and the precision of the obtained motion vectors. The new VLSI architectures that are proposed allow the design of array processors based on the FSBM that can be implemented in FPL devices. These architectures are based on a highly efficient core, that combines both pipelining and parallel processing techniques to design powerful motion estimators [4]. This original core is used to derive simpler structures with reduced hardware requirements. The reduction of the complexity of these architectures is achieved by decreasing the precision of the pixel values and/or the spatial resolutions in the current frame, while maintaining the original resolution in the search space. The pixel precision is configured by defining the number of bits used to represent the input data and by masking or truncating the corresponding Least Significant Bits (LSBs). The spacial resolution is adjusted by sub-sampling the blocks of the current frame. By doing so, the best candidate block in the previous frame is still exhaustively searched but the SAD of each candidate block is computed by using only sparse pixels. Consider a typical setup with 16×16 pixels blocks. By applying $2 : 1$ or $4 : 1$ alternate sub-sampling schemes the number of considered pixels decreases by $1/4$ and $1/16$, respectively.

The efficiency of the proposed structures was evaluated by implementing these customizable core-based architectures in Field Programmable Gate

Arrays (FPGA). It is shown that the amount of hardware required when sub-sampling and truncation techniques are applied is considerably reduced. This fact allows the usage of a common framework for designing a wide range of motion estimation processors with different characteristics that fit well in current FPGAs, being a real alternative to those fast motion estimation techniques that apply non-regular processing. Moreover, experimental results obtained with benchmark video sequences show that the application of these techniques does not introduce a significant degradation in the quality of the coded video sequences.

5.2   Base FSBM Architecture

Several FSBM structures have been proposed over the last few years (e.g.: [5, 6, 4]). Very recently, a new class of parameterizable hardware architectures that is characterized by offering minimum latency, maximum throughput and a full and efficient utilization of the hardware resources was presented in [4]. This last characteristic is a fundamental requisite in any FPL system, due to the limited amount of hardware resources. To achieve such performance levels, a peculiar and innovative processing scheme, based on a cylindrical hardware structure and on the zig-zag processing sequence proposed by Vos [6], was adopted (see Fig. 5.1). With such a scheme, not only is it possible to minimize the processing time, but it also provides the ability to prevent the usage of some

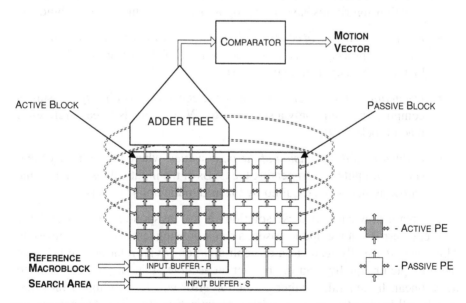

Figure 5.1.   Processor array proposed in [4] based on an innovative cylindrical structure and adopting the zig-zag processing scheme proposed by Vos [6] ($N = 4$, $p = 2$).

hardware structures—the so called *passive processor elements (PEs)*, that do not carry useful information in many clock cycles.

Besides this set of performance and implementation characteristics, this class of processors features also a scalable and configurable architecture, making it possible to easily adapt the processor configuration to fulfill the requisites of a given video coder. By adjusting a small set of implementation parameters, processing structures with distinct performance and hardware requirements are obtained, providing the ability to adjust the required hardware resources to the target implementation technology. While high performance processors, that inevitably require more resources, are more suited for implementations using technologies such as ASIC or Sea-of-Gates, those low-cost processors, that are meant to be implemented in FPL devices, with limited hardware resources, should use configurations requiring reduced amounts of hardware.

5.3   Architectures for Limited Resources Devices

Despite the set of configurable properties offered by FSBM architectures and, in particular, by the class of processors proposed in [4], restricted hardware technologies, such as FPL devices, often do not provide enough hardware resources to implement such processors. In such cases, sub-optimal motion estimation algorithms are usually adopted, which provide faster processing and require reduced amounts of hardware. Different categories of sub-optimal motion estimation algorithms have been proposed, based on three main techniques:

- *Reduction of the set of considered candidate motion vectors*, by restricting the search procedure in the previous frame to a given search pattern using hierarchical search strategies [2, 3];

- *Decimation at the pixel level*, where the considered similarity measure is computed by using only a subset of the $N \times N$ pixels of each reference macroblock [7–9];

- *Reduction of the precision of the pixel values*, where the similarity measure is computed by truncating the LSBs of the input values to reduce the hardware resources required by the arithmetic units [10, 9].

The main drawback of these solutions is a corresponding increase of the prediction error that inevitably arises as result of using less accurate estimation. This tradeoff usually leads to a difficult and non-trivial relationship between the final picture quality and the prediction accuracy that can not be assumed to be linear. In general, a larger prediction error will lead to higher bit rates, which will lead to the usage of greater quantization step sizes to compensate this increase, thus affecting the quality of the decoded images.

Up until now only a few VLSI architectures have been proposed to implement fast motion estimation algorithms, by restricting the search positions according to a given search pattern [9]. In general, they imply the usage of non-regular processing structures and require higher control overheads, which complicates the design of efficient systolic structures. Consequently, they have been extensively used in software applications, where such restrictions do not usually apply so strictly.

The set of architectures proposed here try to combine the advantages offered by the regular and efficient FSBM structures proposed in [4] with several strategies to reduce the amount of hardware that are usually offered by sub-optimal motion estimation algorithms. To implement such architectures on FPL devices, such as FPGAs, the original FSBM architecture will be adapted to apply two of the three decimation categories described above: decimation at the pixel level and reduction of the precision of the pixel values.

5.3.1   Decimation at the Pixel Level

By applying decimation at the pixel level, the image data of the current frame is sub-sampled, by considering alternate pixels in each orthogonal direction. This scheme corresponds to using a lower resolution version of the reference frame in the search procedure that is carried out within the previous full-resolution frame. The SAD measure for a configuration using a $2^S : 1$ sub-sampling in each direction is given by (considering N a power of 2):

$$SAD(l, c) = \sum_{i=0}^{\frac{N}{2^S}-1} \sum_{j=0}^{\frac{N}{2^S}-1} \left| x_t(i.2^S, j.2^S) - x_{t-1}(l + i.2^S, c + j.2^S) \right| \quad (5.1)$$

The FSBM circuit proposed in [4] can be easily adapted to carry out this type of sub-sampling. In fact, considering that the computation of the SAD similarity measure is performed in the active block of the processor, the decimation can be implemented by replacing the corresponding set of active PEs by passive PEs. By doing so, only the pixels with coordinates $(i.2^S, j.2^S)$ will be considered and significant amounts of hardware resources can be saved.

Two different decimation patterns can be applied to the active block of the processor: *i)* an *aligned pattern*, in which the active PEs are aligned in rows and columns, as shown in Fig. 5.2a) and *ii)* a *checkerboard pattern*, in which the active PEs are placed as it is shown in Fig. 5.2b). Although this latter pattern may seem to provide better estimations, since it reduces the correlation degree of the pixel values, experimental results have shown that the usage of this pattern may cause a Peak Signal to Noise Ratio (PSNR) reduction of the coded video of about 0.5–1.0 dB. On the other hand, by adopting the sub-sampling pattern presented in Fig. 5.2a) the degradation of the PSNR is

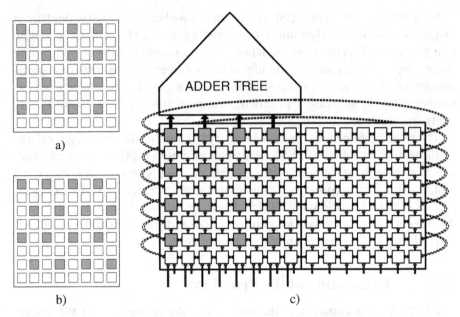

Figure 5.2. *a*) Aligned pattern. *b*) Checkerboard pattern. *c*) Modified processing array to carry out a 2:1 decimation function using the architecture proposed in [4] ($N = 8$, $p = 4$).

negligible and significant amounts of hardware resources can be saved. In fact, not only do the passive PEs require considerably fewer hardware resources, but also the number of inputs of the adder tree block is reduced by the sub-sampling factor (S), which further reduces the amount of required hardware (see Fig. 5.2).

According to the classical signal processing theory, this pixel decimation approach can be seen as a reduction of the sampling frequency of the image. Hence, it would be reasonable to expect that the PSNR of the coded video could be improved if anti-aliasing filtering were applied to the search or/and reference areas of the image. However, experimental results showed that the usage of such filters causes the opposite effect. Whether the filter is only applied to the search area, to the reference area or to both areas, the PSNR of the coded video always suffers a reduction of 0.5–5.0 dB, depending on the type of video sequence, i.e., on the amount of movement. Furthermore, since this PSNR reduction was obtained using both 3×3 and 5×5 Gaussian filters, it can be concluded that the PSNR reduction is independent of the filter type. Moreover, significant hardware resources would have to be wasted in the implementation of these 2D filters. As a result, the pixel decimation approach was implemented using the pattern depicted in Fig. 5.2a) and not using any anti-aliasing filtering. The block diagram of the modified processing array is presented in Fig. 5.2c).

5.3.2 Reduction of the Precision of the Pixel Values

Another strategy to decrease the amount of hardware required by FSBM processors is to reduce the bit resolution of the pixel values considered in the computation of the SAD similarity function by truncating the LSBs. By adopting this strategy alone ($S = 0$) or in conjunction with the previously described sub-sampling method ($0 < S < \log_2 N$), the SAD measure is given by:

$$SAD(l, c) = \sum_{i=0}^{\frac{N}{2^S}-1} \sum_{j=0}^{\frac{N}{2^S}-1} \left| \left\lfloor \frac{x_t(i.2^S, j.2^S)}{2^T} \right\rfloor - \left\lfloor \frac{x_{t-1}(l + i.2^S, c + j.2^S)}{2^T} \right\rfloor \right| \tag{5.2}$$

$$\equiv \sum_{i=0}^{\frac{N}{2^S}-1} \sum_{j=0}^{\frac{N}{2^S}-1} \left| x_t(i.2^S, j.2^S)_{7:T} - x_{t-1}(l + i.2^S, c + j.2^S)_{7:T} \right| \tag{5.3}$$

where T is the number of truncated bits and $x_t(i, j)_{7:T}$ are the $(8 - T)$ most significant bits of a pixel value of the tth frame.

The adaptation of the original FSBM architecture [4] to apply this bit truncation scheme is straightforward: it is only necessary to reduce the operand widths of the several arithmetic units implemented in the active PEs, in the adder tree blocks and in the comparator circuits. Such modifications potentially increase the maximum frequency of the pipeline and will significantly reduce the amount of required hardware, thus providing the conditions that will make it possible to implement the motion estimation processors in FPL devices.

5.4 Implementation and Experimental Results

Several different setups of the proposed customizable core-based architectures for motion estimation were synthesized and implemented in a general proposed VIRTEX XCV3200E-7 FPGA using the Xilinx Synthesis Tool from ISE 5.2.1. The considered set of configurations assumed each macroblock composed by 16×16 pixels ($N = 16$) and a maximum displacement in each direction of the search area of $p = 16$ pixels. These configurations were thoroughly tested using sub-sampling factors (S) varying between 0 and 2 and a number of truncated bits (T) of 0, 2 and 4. The experimental results of these implementations are presented in Tables 5.1 and 5.3.

The proposed core-based architectures can provide significant savings in the required hardware resources. From the set of configurations presented in Table 5.1, one can observe that reduction factors of about 75% can be obtained by using a sub-sampling factor $S = 2$ and by truncating the 4 LSBs. However, this relation should not be assumed to be linear. By considering only the pixel

Table 5.1. Percentage of the CLB slices and LUTs that are required to implement each configuration of the proposed core-based architecture for fast motion estimation in a VIRTEX XCV3200E-7 FPGA ($N = 16$; $p = 16$).

S	$T = 0$		$T = 2$		$T = 4$	
	CLB Slices	LUTs	CLB Slices	LUTs	CLB Slices	LUTs
0	90.7%	30.7%	62.6%	23.5%	43.8%	16.3%
1	56.2%	19.0%	38.2%	14.6%	26.6%	10.7%
2	44.7%	14.8%	30.1%	11.4%	21.0%	7.8%

level decimation mechanism ($T = 0$), it can be shown that a reduction of about 38% is obtained by using a $2 : 1$ sub-sampling factor ($S = 1$), while a $4 : 1$ decimation will provide a reduction of about 51%. The same observation can be made by considering only the reduction of the precision of the pixel values ($S = 0$). While using 6 representation bits ($T = 2$) a reduction of the number of CLB slices of about 31% is obtained (a reduction of about 23% of the number of Look-up Tables (LUTs)). If only 4 representation bits are considered ($T = 4$) a reduction of about 51% of the CLB slices is achieved (a reduction of about 47% of the number of LUTs). Table 5.2 presents the set of FPGA devices that should be used by each configuration in order to maximize the efficiency of the hardware resources used by each processor.

Table 5.3 presents the variation of the maximum operating frequency of the considered configurations with the number of truncated bits (T). Contrary to what could be expected, the reduction of the operands width of the several arithmetic units does not significantly influence the processors performance. This fact can be explained if one takes into account the synthesis mechanism

Table 5.2. Alternative FPGA devices to implement each of the considered configurations ($N = 16$; $p = 16$).

S	$T = 0$	$T = 2$	$T = 4$
0	XCV3200	XCV2600	XCV1600
1	XCV2000	XCV1600	XCV1000
2	XCV1600	XCV1000	XCV600

Table 5.3. Variation of the maximum operating frequency with the number of truncated bits (T) ($N = 16$; $p = 16$).

$T = 0$	$T = 2$	$T = 4$
76.1 MHz	77.8 MHz	79.8 MHz

that is used by this family of FPGAs to synthesize and map the logic circuits using built in fast carry logic and LUTs.

To assess and evaluate the efficiency of the synthesized processors in an implementation based on FPL devices, they were embedded as motion estimation co-processors in the H.263 video encoder provided by Telenor R&D [11], by transferring the estimated motion vectors to the video coder. Peak signal-to-noise ratio (PSNR) and bit-rate measures were used to evaluate the performance of each architecture. These results were also compared with those obtained with a sub-optimal 4-step logarithmic search algorithm [1], implemented in software.

The first 300 frames of several QCIF benchmark video sequences with different spatial detail and amount of movement were coded in interframe mode, by considering a GOP length of 30 frames and a quantization step with intermediate size of $\Delta = 30$ to keep the quantization error as constant as possible.

Fig. 5.3 presents the PSNR values obtained for the *carphone* and *mobile* video sequences, characterized by the presence of high amounts of movement and high spatial detail, respectively. Several different setups in what concerns the number of truncated bits (T) and the sub-sampling factor (S) for the decimation at the pixel level were considered. The PSNR value obtained for the INTER type frames of the *mobile* sequence is almost constant (25–26 dB), which demonstrates that the degradation introduced by using the reduced hardware architectures is negligible: less than 0.15 dB when the 4 LSBs are truncated and the sub-sampling factor is 2 : 1. In contrast, it can be seen that the PSNR values obtained for the *carphone* sequence varies significantly along the time. The main reason for this fact is the amount of movement present in this sequence that is also varying. Even so, the proposed reduced hardware architectures only introduce a slight degradation in the quality of the coded frames, when compared with the performances of both the reference FSBM architecture $(S = T = 0)$ and of the 4-steps logarithmic search algorithm. A maximum reduction of about 0.5 dB is observed in a few frames when the PSNR value obtained with the original (reference) architecture is greater than 30 dB.

As it was previously noted, these video quality results were obtained with a constant quantization step size. The observed small decrease of the PSNR can be explained by the slight increase of the quantization error, as a consequence of the inherent increase of the prediction differences in the motion compensated block, obtained with these sub-optimal matching processors. The required bit-rate to store or transfer the two video sequences considered above and two other additional sequences (*Miss America* and *Silent*) is presented in Table 5.4. The relative values presented in Table 5.4 demonstrate the corresponding increment of the average bit-rate when the number of truncated bits and the sub-sampling factor increase. This increment reaches its maximum value of 12% for the *carphone* sequence and for a processor setup with only 4 representation

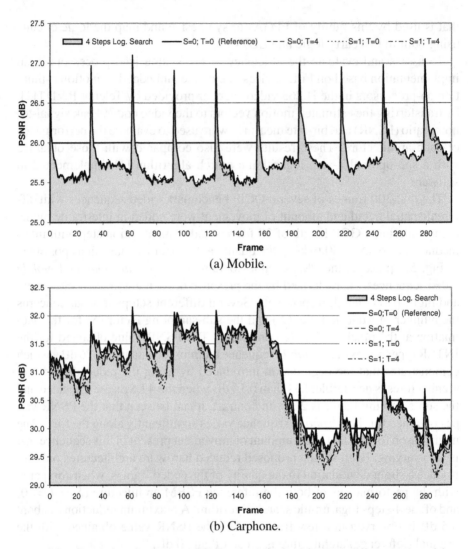

Figure 5.3. Comparison between the PSNR values obtained for three different setups of the proposed reduced hardware architecture, for the original reference FSBM configuration and for the 4-steps logarithmic search algorithm.

bits and a pixel decimation of 2 : 1, due to the presence of a lot of movement. Nevertheless, the values obtained for the remaining three video sequences, with less amount of movement, are quite smaller and similar to those ones obtained with the 4-steps logarithmic search algorithm.

Consequently, from these results it can be concluded that the configuration using a 2 : 1 decimation ($S = 1$) and 4 representation bits ($T = 4$) presents the best tradeoff between hardware cost (a reduction of about 70%) and video quality. Moreover, this configuration can be implemented by using the

Table 5.4. Variation of the output bit rate to encode the considered video sequences by using different setups of the proposed core-base architecture and the 4-steps logarithmic search algorithm.

(a) Miss America.   (b) Silent.

S	$T = 0$	$T = 2$	$T = 4$	S	$T = 0$	$T = 2$	$T = 4$
0	31.9 kbps	−0.8%	+3.4%	0	49.9 kbps	+0.0%	+2.7%
1	+1.7%	+3.0%	+3.8%	1	+4.5%	+4.1%	+8.0%
2	+3.9%	+3.8%	+3.8%	2	+10.7%	+8.4%	+8.6%
Four-step logarithmic search			+0.3%	Four-step logarithmic search			+0.8%

(c) Mobile.   (d) Carphone.

S	$T = 0$	$T = 2$	$T = 4$	S	$T = 0$	$T = 2$	$T = 4$
0	271.4 kbps	+0.0%	+0.3%	0	79.3 kbps	−0.8%	+4.8%
1	+1.1%	+0.3%	+0.5%	1	+7.3%	+5.8%	+11.6%
2	+6.7%	+0.6%	+0.5%	2	+14.5%	+11.6%	+12.0%
Four-step logarithmic search			−0.1%	Four-step logarithmic search			+1.7%

lower-cost XCV1000 FPGA, which has only about 40% of the total number of system gates provided by the XCV3200 FPGA.

5.5   Conclusion

New customizable core-based architectures are proposed herein to implement real-time motion estimation processors on FPL devices, such as FPGAs. The base core of these architectures is a new 2-D array structure for FSBM motion estimation that leads to an efficient usage of hardware resources, which is a fundamental requisite in any FPL based system. The proposed architectures consist of a wide range of processing structures based on the FSBM algorithm with different hardware requirements. The reduction in the amount of required hardware is achieved by applying decimation at the pixel and quantization levels, but still searching all candidate blocks of a given search area.

The proposed core-based architectures were implemented on FPGA devices from Xilinx and their performance was evaluated by including the motion estimation processors on a complete video encoding system. Experimental results were obtained by sub-sampling the block of the current frame with 2 : 1 and 4 : 1 decimation factors and by truncating 2 or 4 LSBs of the representation. From the results obtained it can be concluded that a significant reduction of the required hardware resources can be achieved with these architectures. Moreover, neither the quality of the coded video is compromised nor the corresponding bit-rate is significantly increased. One can also conclude from the obtained results that the configuration using a $2:1$ decimation ($S = 1$) and 4 representation bits ($T = 4$) presents the best tradeoff between hardware cost (a reduction of about 70%) and video quality.

Acknowledgments

This work has been supported by the POSI program and the *Portuguese Foundation for Science and for Technology* (FCT) under the research project *Configurable and Optimized Processing Structures for Motion Estimation* (COSME) POSI/CHS/40877/2001.

References

[1] Vasudev Bhaskaran and Konstantinos Konstantinides, *Image and Video Compression Standards: Algorithms and Architectures*, Kluwer Academic Publishers, second edition, June 1997.

[2] T. Koga, K. Iinuma, A. Hirano, Y. Iijima and T. Ishiguro, "Motion-compensated interframe coding for video conferencing," in *Proc. Nat. Telecomm. Conference*, New Orleans, LA, November 1981, pp. G5.3.1–G5.3.5.

[3] J. R. Jain and A. K. Jain, "Displacement measurement and its application in interframe image coding," *IEEE Transactions on Communications*, volume COM-29, no. 12, pp. 1799–1808, December 1981.

[4] Nuno Roma and Leonel Sousa, "Efficient and configurable full search block matching processors," *IEEE Transactions on Circuits and Systems for Video Technology*, volume 12, no. 12, pp. 1160–1167, December 2002.

[5] Y. Ooi, "Motion estimation system design," in *Digital Signal Processing for Multimedia Systems* (edited by Keshab K. Parhi and Takao Nishitani), Marcel Dekker, Inc, chapter 12, pp. 299–327, 1999.

[6] L. Vos and M. Stegherr, "Parameterizable VLSI architectures for the full-search block-matching algorithm," *IEEE Transactions on Circuits and Systems*, volume 36, no. 10, pp. 1309–1316, October 1989.

[7] B. Liu and A. Zaccarin, "New fast algorithms for the estimation of block matching vectors," *IEEE Transactions on Circuits and Systems for Video Technology*, volume 3, no. 2, pp. 148–157, April 1993.

[8] E. Ogura, Y. Ikenaga, Y. Iida, Y. Hosoya, M. Takashima and K. Yamash, "A cost effective motion estimation processor LSI using a simple and efficient algorithm," in *Proceedings of International Conference on Consumer Electronics—ICCE*, 1995, pp. 248–249.

[9] Seongsoo Lee, Jeong-Min Kim, and Soo-Ik Chae, "New motion estimation algorithm using adaptively-quantized low bit resolution image and its vlsi architecture for MPEG2 video coding," *IEEE Transactions on Circuits and Systems for Video Technology*, volume 8, no. 6, pp. 734–744, October 1998.

[10] Z. L. He, K. K. Chan, C. Y. Tsui and M. L. Liou, "Low power motion estimation design using adaptative pixel truncation," in *Proceedings of the 1997 international symposium on Low power electronics and design*, Monterey—USA, August 1997, pp. 167–171.

[11] Telenor, *TMN (Test Model Near Term)—(H.263) encoder/decoder—version 2.0—source code*, Telenor Research and Development, Norway, June 1996.

Methodologies and Tools

Chapter 6

Enabling Run-time Task Relocation on Reconfigurable Systems

J-Y. Mignolet, V. Nollet, P. Coene, D. Verkest[1,2],
S. Vernalde, R. Lauwereins[2]
IMEC vzw, Kapeldreef 75, 3001 Leuven, BELGIUM
{mignolet, nollet, coene}imec.be

[1] *also Professor at Vrije Universiteit Brussel*

[2] *also Professor at Katholieke Universiteit Leuven*

Abstract The ability to (re)schedule a task either in hardware or software will be an important asset in a reconfigurable systems-on-chip. To support this feature we have developed an infrastructure that, combined with a suitable design environment permits the implementation and management of hardware/software relocatable tasks. This paper presents the general scope of our research, and details the communication scheme, the design environment and the hardware/software context switching issues. The infrastructure proved its feasibility by allowing us to design a relocatable video decoder. When implemented on an embedded platform, the decoder performs at 23 frames/s (320×240 pixels, 16 bits per pixel) in reconfigurable hardware and 6 frames/s in software.

Keywords: Run-time task relocation, reconfigurable systems, operating system, network-on-chip

Introduction

Today, emerging run-time reconfigurable hardware solutions are offering new perspectives on the use of hardware accelerators. Indeed, a piece of reconfigurable hardware can now be used to run different tasks in a sequential way. By using an adequate operating system, software-like tasks can be created, deleted and pre-empted in hardware as it is done in software.

P. Lysaght and W. Rosenstiel (eds.),
New Algorithms, Architectures and Applications for Reconfigurable Computing, 69–80.
© 2005 *Springer. Printed in the Netherlands.*

A platform composed of a set of these reconfigurable hardware blocks and of instruction-set processors (ISP) can be used to combine two important assets: flexibility (of software) and performance (of hardware). An operating system can manage the different tasks of an application and spawn them in hardware or in software, depending on their computational requirements and on the quality of service that the user expects from these applications.

Design methodology for applications that can be relocated from hardware to software and vice-versa is a challenging research topic related to these platforms. The application should be developed in a way that ensures an equivalent behavior for its hardware and software implementations to allow run-time relocation. Furthermore, equivalence of states between hardware and software should be studied to efficiently enable heterogeneous context switches.

In the scope of our research on a general-purpose programmable platform based on reconfigurable hardware, we have developed an infrastructure for the design and management of relocatable tasks. The combination of a uniform communication scheme and OCAPI-xl [8, 9], a C++ library for unified hardware/software system design, allowed us to develop a relocatable video decoder. This was demonstrated on a platform composed of a commercial FPGA and a general purpose ISP. It is the first time to our knowledge that full hardware/software multitasking is addressed, in such a way that the operating system is able to spawn and relocate a task either in hardware or software.

The remainder of this paper is organized as follows. Section 6.1 puts the problem into perspective by positioning it in our general research activity. Section 6.2 describes the communication scheme we developed on the platform and its impact on the task management. Section 6.3 presents the object oriented design environment we used to design the application. Section 6.4 discusses the heterogeneous context switching issues. Section 6.5 gives an overview of implementation results on a specific case study. Finally some conclusions are drawn in Section 6.6. Related work [3, 5–7, 10, 12] will be discussed throughout the paper.

6.1 Hardware/Software Multitasking on a Reconfigurable Computing Platform

The problem of designing and managing relocatable tasks fits into the more general research topic of hardware/software multitasking on a reconfigurable computing platform for networked portable multimedia appliances. The aim is to increase the computation power of current multimedia portable devices (such as personal digital assistants or mobile phones) while keeping their flexibility. Performance should be coupled with low power consumption, since portable devices are battery-operated. Flexibility is required because different applications

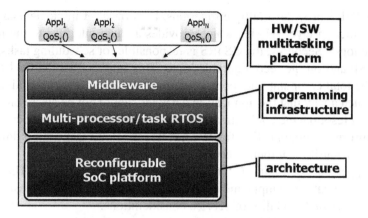

Figure 6.1. Our research activity.

will run on the device, with different architecture requirements. Moreover, it enables upgrading and downloading of new applications. Reconfigurable hardware meets these two requirements and is therefore a valid solution to this problem.

Our research activity addresses different parts of the problem, as shown in Figure 6.1. A complete description is presented in [1].

The bottom part represents the platform activity, which consists in defining suitable architectures for reconfigurable computing platforms. The selection of the correct granularity for the reconfigurable hardware blocks and the development of the interconnection network that will handle the communication between the different parts of the system are two of the challenges for this activity.

The interconnection network plays an important role in our infrastructure, since it supports the communication of the system. Networks-on-chip provide a solution for handling communication in complex systems-on-chip (SoC). We are studying packet-switched interconnection networks for reconfigurable platforms [2]. To assist this research, we develop "soft" interconnection networks on commercial reconfigurable hardware. They are qualified soft because they are implemented using the reconfigurable fabric, while future platforms will use fixed networks implemented using standard ASIC technology. This soft interconnection network divides the reconfigurable hardware in tiles of equal size. Every tile can run one task at a given moment.

The middle part of Figure 6.1 represents the operating system for reconfigurable systems (OS4RS) we have developed to manage the tasks over the different resources. In order to handle hardware tasks, we have developed extensions as a complement to the traditional operating system.

The OS4RS provides multiple functions. First of all, it implements a hardware abstraction layer (HAL), which provides a clean interface to the reconfigurable logic. Secondly, the OS4RS is responsible for scheduling tasks, both on the ISP and on the reconfigurable logic. This implies that the OS4RS abstracts the total computational pool, containing the ISP and the reconfigurable tiles, in such a way that the application designer should not be aware on which computing resource the application will run. A critical part of the functionality is the uniform communication framework, which allows tasks to send/receive messages, regardless of their execution location.

The upper part of Figure 6.1 represents the middleware layer. This layer takes the application as input and decides on the partitioning of the tasks. This decision is driven by quality-of-service considerations.

The application should be designed in such a way that it can be executed on the platform. In a first approach, we use a uniform HW/SW design environment to design the application. Although it ensures a common behavior for both HW and SW version of the task, it still requires both versions of the task to be present in memory. In future work, we will look at unified code that can be interpreted by the middleware layer and spawned either in HW or SW. This approach will not only be platform independent similar to JAVA, it will also reduce the memory footprint, since the software and the hardware code will be integrated.

6.2 Uniform Communication Scheme

Relocating a task from hardware to software should not affect the way other tasks are communicating with the relocated task. By providing a uniform communication scheme for hardware and software tasks, the OS4RS we developed hides this complexity.

In our approach, inter-task communication is based on message passing. Messages are transferred from one task to another in a common format for both hardware and software tasks. Both the operating system and the hardware architecture should therefore support this kind of communication.

Every task is assigned a logical address. Whenever the OS4RS schedules a task in hardware, an address translation table is updated. This address translation table allows the operating system to translate a logical address into a physical address and vice versa. The assigned physical address is based on the location of the task in the interconnection network (ICN).

The OS4RS provides a message passing API, which uses these logical/physical addresses to route the messages. In our communication scheme, three subtypes of message passing between tasks can be distinguished (Figure 6.2).

Messages between two tasks, both scheduled on the ISP (P1 and P2), are routed solely based on their logical address and do not pass the HAL.

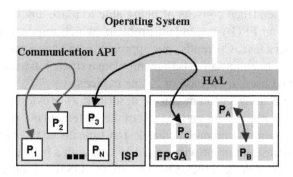

Figure 6.2. Message passing between tasks.

Communication between an ISP task and a FPGA task (P3 and Pc) does pass through the hardware abstraction layer. In this case, a translation between the logical address and the physical address is performed by the communication API. The task's physical address allows the HAL to determine on which tile of the ICN the sending or receiving task is executing.

On the hardware side, the packet-switched interconnection network is providing the necessary support for message passing. Messages between tasks, both scheduled in hardware, are routed inside the interconnection network without passing through the HAL. Nevertheless, since the operating system controls the task placement, it also controls the way the messages are routed inside the ICN, by adjusting the hardware task routing tables.

The packet-switched interconnection network, which supports the hardware communication in our infrastructure, solves some operating system issues related to hardware management such as task placement, location independence, routing, and inter-task communication. Diessel and Wigley previously listed these issues in [3].

Task placement is the problem of positioning a task somewhere in the reconfigurable hardware fabric. At design time, task placement is realized by using place and route tools from the reconfigurable hardware vendor. This usually generates an irregular task footprint. At run-time, the management software is responsible for arranging all the tasks inside the reconfigurable fabric. When using irregular task shapes, the management software needs to run a complex fitting algorithm (e.g. [6, 7]). Executing this placement algorithm considerably increases run-time overhead. In our infrastructure, the designer constrains the place and route tool to fit the task in the shape of a tile. Run-time task placement is therefore greatly facilitated, since every tile has the same size and same shape. The OS4RS is aware of the tile usage at any moment. As a consequence, it can spawn a new task without placement overhead by replacing the tile content through partial reconfiguration of the FPGA.

Location independence consists of being able to place any task in any free location. This is an FPGA-dependent problem, which requires a relocatable bitstream for every task. Currently, our approach is to have a partial bitstream for every tile. A better alternative is to manipulate a single bitstream at run-time (Jbits [4] could be used in the case of Xilinx devices).

The run-time routing problem can be described as providing connectivity between the newly placed task and the rest of the system. In our case, a communication infrastructure is implemented at design-time inside the interconnection network. This infrastructure provides the new task with a fixed communication interface, based on routing tables. Once again, the OS4RS should not run any complex algorithm. Its only action is updating the routing tables every time a new task is inserted/removed from the reconfigurable hardware.

The issue of inter-task communication is handled by the OS4RS, as described earlier this section.

Our architecture makes a trade-off between area and run-time overhead. As every tile is identical in size and shape, the area fragmentation (as defined by Wigley and Kearney in [5]) is indeed higher than in a system where the logic blocks can have different sizes and shapes. However, the OS4RS will only need a very small execution time to spawn a task on the reconfigurable hardware, since the allocation algorithm is limited to the check of tile availability.

6.3 Unified Design of Hardware and Software with OCAPI-xl

A challenging step in the design of relocatable tasks is to provide a common behavior for the HW and the SW implementation of a task. One possibility to achieve this is to use a unified representation that can be refined to both hardware and software.

OCAPI-xl [8, 9] provides this ability. OCAPI-xl is a C++ library that allows unified hardware/software system design. Through the use of the set of objects from OCAPI-xl, a designer can represent the application as communicating threads. The objects contain timing information, allowing cycle-true simulation of the system. Once the system is designed, automatic code generation for both hardware and software is available. This ensures a uniform behavior for both implementations in our heterogeneous reconfigurable system.

Through the use of the FLI (Foreign Language Interface) feature of OCAPI-xl, an interface can be designed that represents the communication with the other tasks. This interface provides functions like send_message and receive_message that will afterwards be expanded to the corresponding hardware or software implementation code. This ensures a communication scheme that is common to both implementations.

6.4 Heterogeneous Context Switch Issues

It is possible for the programmer to know at design time on which of the heterogeneous processors the tasks preferably should run (as described by Lilja in [11]). However, our architecture does not guarantee run-time availability of hardware tiles. Furthermore, the switch latency of hardware tasks (in the range of 20ms on a FPGA) severely limits the number of time-based context switches. We therefore prefer spatial multitasking in hardware, in contrast to the time-based multitasking presented in [10, 12]. Since the number of tiles is limited, the OS4RS is forced to decide at run-time on the allocation of resources, in order to achieve maximum performance. Consequently, it should be possible for the OS4RS to pre-empt and relocate tasks from the reconfigurable logic to the ISP and vice versa.

The ISP registers and the task memory completely describe the state of any task running on the ISP. Consequently, the state of a preempted task can be fully saved by pushing all the ISP registers on the task stack. Whenever the task gets rescheduled at the ISP, simply popping the register values from its stack and initializing the registers with these values restores its state.

This approach is not usable for a hardware task, since it depicts its state in a completely different way: state information is held in several registers, latches and internal memory, in a way that is very specific for a given task implementation. There is no simple, universal state representation, as for tasks executing on the ISP. Nevertheless, the operating system will need a way to extract and restore the state of a task executing in hardware, since this is a key issue when enabling heterogeneous context switches.

A way to extract and restore state when dealing with tasks executing on the reconfigurable logic, is described in [10, 12]. State extraction is achieved by getting all status information bits out of the read back bitstream. This way, manipulation of the configuration bitstream allows re-initializing the hardware task. Adopting this methodology to enable heterogeneous context switches would require a translation layer in the operating system, allowing it to translate an ISP type state into FPGA state bits and vice versa. Furthermore, with this technique, the exact position of all the configuration bits in the bitstream must be known. It is clear that this kind of approach does not produce a universally applicable solution for storing/restoring task state.

We propose to use a high level abstraction of the task state information. This way the OS4RS is able to dynamically reschedule a task from the ISP to the reconfigurable logic and vice versa. This technique a based on an idea presented in [13]. Figure 6.3a represents a relocatable task, containing several states. This task contains 2 switch-point states, at which the operating system can relocate the task. The entire switch process is described in detail by Figure 6.4. In order to relocate a task, the operating system can signal that task at any

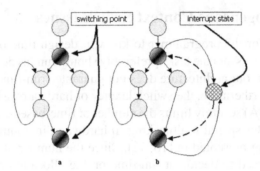

Figure 6.3. Relocatable task.

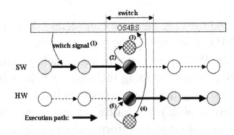

Figure 6.4. Task switching: from software to hardware.

time (1). Whenever the signaled task reaches a switch-point, it goes into the interrupted state (2) (Figure 6.3b). In this interrupted state all the relevant state information of the switch-point is transferred to the OS4RS (3). Consequently, the OS4RS will re-initiate the task on the second heterogeneous processor using the received state information (4). The task resumes on the second processor, by continuing to execute in the corresponding switch-point (5). Note that the task described in Figure 6.3 contains multiple switch-points, which makes it possible that the state information that needs to be transferred to the OS4RS can be different for each switch-point. Furthermore, the unified design of both the ISP and FPGA version of a task, as described in Section 6.3, ensures that the position of the switch-points and the state information are identical.

The relocatable video decoder, described in Section 6.5, illustrates that the developed operating system is able to dynamically reschedule a task from the ISP to the reconfigurable logic and vice versa. At this point in time, this simplified application contains only one switchable state, which contains no state information.

The insertion of these "low overhead" switch-points will also be strongly architecture dependent: in case of a shared memory between the ISP and the reconfigurable logic, transferring state can be as simple as passing a pointer, while in case of distributed memory, data will have to be copied.

On a long term, the design tool should be able to create these switch-points automatically. One of the inputs of the design tool will be the target architecture. The OS4RS will then use these switch-points to perform the context switches in a way hidden from the designer.

6.5 Relocatable Video Decoder

As an illustration of our infrastructure a relocatable video decoder is presented. First the platform on which the decoder was implemented is described. Then the decoder implementation is detailed. Finally performance and implementation results are presented.

6.5.1 The T-ReCS Gecko Demonstrator

Based on the concepts presented in Section 6.1, we have developed a first reconfigurable computing platform for HW/SW multitasking. The Gecko demonstrator (Figure 6.5) is a platform composed of a Compaq iPAQ 3760 and a Xilinx Virtex 2 FPGA. The iPAQ is a personal digital assistant (PDA) that features a StrongARM SA-1110 ISP and an expansion bus that allows connection of an external device. The FPGA is a XC2V6000 containing 6000k system gates.

The FPGA is mounted on a generic prototyping board connected to the iPAQ via the expansion bus. On the FPGA, we developed a soft packet-switched interconnection network composed of two application tiles and one interface tile.

6.5.2 The Video Decoder

Our Gecko platform is showcasing a video decoder that can be executed in hardware or in software and that can be rescheduled at run-time.

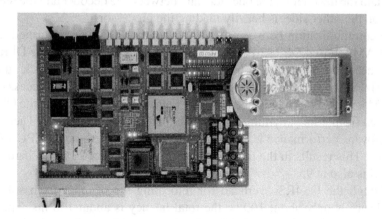

Figure 6.5. The T-ReCS Gecko demonstrator.

The video decoder is a motion JPEG frame decoder. A send thread passes the coded frames one by one to the decoder thread. This thread decodes the frames and sends them, one macroblock at a time, to a receive thread that reconstructs the images and displays them. The send thread and the receive thread run in software on the iPAQ, while the decoder thread can be scheduled in HW or in SW.

The switch point has been inserted at the end of the frame because, at this point, no state information has to be transferred from HW to SW or vice-versa.

6.5.3 Results

Two implementations of the JPEG decoder have been designed. The first one is quality factor and run-length encoding specific (referred as specific hereafter), meaning that the quantization tables and the Huffman tables are fixed, while the second one can accept any of these tables (referred as general hereafter). Both implementations target the 4:2:0 sampling ratio. The results of the implementation of the decoders in hardware are 9570 LUTs for the specific implementation and 15901 LUTs for the general one. (These results are given by the report file from the Synplicity(r) Synplify Pro(tm) advanced FPGA synthesis tool, targeting the Virtex2 XC2V6000 device, speed grade −4, and for a required clock frequency of 40 MHz).

The frame rate of the decoder is 6 frames per second (fps) for the software implementation and 23 fps for the hardware. These results are the same for both general and specific implementation. The clock runs at 40 MHz, which is the maximum frequency that can be used for this application on the FPGA. When achieving 6 fps in software, the CPU load is about 95%. Moving the task to hardware reduces the computational load of the CPU, but increases the load generated by the communication. Indeed, the communication between the send thread and the decoder on the one side, and between the decoder and the receive thread on the other side, is heavily loading the processor.

The communication between the iPAQ and the FPGA is performed using BlockRAM internal DPRAMs of the Xilinx Virtex FPGA. While the DPRAM can be accessed at about 20 MHz, the CPU memory access clock runs at 103 MHz. Since the CPU is using a synchronous RAM scheme to access these DPRAMs, wait-states have to be inserted. During these wait-states, the CPU is prevented from doing anything else, which increases the CPU load. Therefore, the hardware performance is mainly limited by the speed of the CPU-FPGA interface. This results in the fact that for a performance of 23 fps in hardware, the CPU is also at 95.

Although the OS4RS overhead for relocating the decoder from software to hardware is only about 100 ţs, the total latency is about 108 ms. The low OS4RS overhead can be explained by the absence of a complex task placement algorithm. Most of the relocation latency is caused by the actual partial reconfiguration through the slow CPU-FPGA interface. In theory, the total software to hardware relocation latency can be reduced to about 11ms, when performing the partial reconfiguration at full speed. When relocating a task from hardware to software, the total relocation latency is equal to the OS4RS overhead, since in this case no partial reconfiguration is required.

Regarding power dissipation, the demo setup cannot show relevant results. Indeed, the present platform uses an FPGA as reconfigurable hardware. Traditionally, FPGAs are used for prototyping and are not meant to be power efficient. The final platform we are targeting will be composed of new, low-power fine- and coarse-grain reconfigurable hardware that will improve the total power dissipation of the platform. Power efficiency will be provided by the ability of spawning highly parallel, computation intensive tasks on this kind of hardware.

6.6   Conclusions

This paper describes a novel infrastructure for the design and management of relocatable tasks in a reconfigurable SoC. The infrastructure consists of a unified HW/SW communication scheme and a common HW/SW behavior. The uniform communication is ensured by a common message-passing scheme inside the operating system and a packet switched interconnection network. The common behavior is guaranteed by use of a design environment for unified HW/SW system design. The design methodology has been applied to a video decoder implemented on an embedded platform composed of an instruction-set processor and a network-on-FPGA. The video decoder is relocatable and can perform 6 fps in software and 23 fps in hardware. Future work includes automated switch-point placement and implementation in order to have a low context switch overhead when heterogeneously rescheduling tasks.

Acknowledgments

We would like to thank Kamesh Rao of Xilinx for carefully reviewing and commenting this paper.

Part of this research has been funded by the European Commission through the IST-AMDREL project (IST-2001-34379) and by Xilinx Labs, Xilinx Inc. R&D group.

References

[1] J-Y. Mignolet, S. Vernalde, D. Verkest, R. Lauwereins: Enabling hardware-software multitasking on a reconfigurable computing platform for networked portable multimedia appliances. In *Proceedings*

of the International Conference on Engineering Reconfigurable Systems and Architecture 2002, pages 116–122, Las Vegas, June 2002.

[2] Marescaux, T., Bartic, A., Verkest, D., Vernalde, S., and Lauwereins, R.: Interconnection Networks Enable Fine-Grain Dynamic Multi-Tasking on FPGAs. In *Proceedings of the 12th International Conference on Field-Programmable Logic and Applications (FPL'2002)*, pages 795–805, Montpellier France.

[3] Diessel, O., Wigley, G.: Opportunities for Operating Systems Research in Reconfigurable Computing. Technical report ACRC-99–018, Advanced Computing Research Centre, School of Computer and Information Science, University of South Australia, August, 1999.

[4] Guccione, S., Levi, D., Sundararajan, P.: JBits: A Java-based Interface for Reconfigurable Computing, 2nd Annual Military and Aerospace Applications of Programmable Devices and Technologies Conference (MAPLD).

[5] G. Wigley, D. Kearney: The Management of Applications for Reconfigurable Computing using an Operating System. In *Proc. Seventh Asia-Pacific Computer Systems Architecture Conference*, January 2002, ACS Press.

[6] J. Burns, A. Donlin, J. Hogg, S. Singh, M. de Wit: A Dynamic Reconfiguration Run-Time System. In *Proceedings of the 5th IEEE Symposium on FPGA-Based Custom Computing Machines (FCCM '97)*, Napa Valley, CA, April 1997.

[7] H. Walder, M. Platzner: Non-preemptive Multitasking on FPGAs: Task Placement and Footprint Transform. In *Proceedings of the International Conference on Engineering Reconfigurable Systems and Architecture 2002*, pages 24–30, Las Vegas, June 2002.

[8] www.imec.be/ocapi

[9] G. Vanmeerbeeck, P. Schaumont, S. Vernalde, M. Engels, I. Bolsens: Hardware/Software Partitioning of embedded system in OCAPI-xl. In *Proc. CODES'01*, Copenhagen, Denmark, April 2001.

[10] H. Simmler, L. Levinson, R. Männer: Multitasking on FPGA Coprocessors. In *Proc. 10th Int'l Conf. Field Programmable Logic and Applications*, pages 121–130, Villach, Austria, August 2000.

[11] D. Lilja: Partitioning Tasks Between a Pair of Interconnected Heterogeneous Processors: A Case Study. *Concurrency: Practice and Experience*, Vol. 7, No. 3, May 1995, pp. 209–223.

[12] L. Levinson, R. Männer, M. Sesler, H. Simmler: Preemptive Multitasking on FPGAs. In *Proceedings of the 2000 IEEE Symposium on Field Programmable Custom Computing Machines*, Napa Valley, CA, USA, April 2000.

[13] F. Vermeulen, F. Catthoor, L. Nachtergaele, D. Verkest, H. De Man: Power-efficient flexible processor architecture for embedded applications. *IEEE Transactions on Very Large Scale Integration Systems*, Vol. 11, No. 3, June 2003, pp. 376–385.

Chapter 7

A Unified Codesign Environment
For The UltraSONIC Reconfigurable Computer

Theerayod Wiangtong[1], Peter Y.K Cheung[1], Wayne Luk[2]

[1] *Department of Electrical & Electronic Engineering,*
Imperial College, London, UK
{tw1,p.cheung}@imperial.ac.uk

[2] *Department of Computing, Imperial College, London, UK*
wl@imperial.ac.uk

Abstract   This paper presents a codesign environment for the UltraSONIC reconfigurable computing platform which is designed specifically for real-time video applications. A codesign environment with automatic partitioning and scheduling between a host processor and a number of reconfigurable coprocessors is described. A unified runtime environment for both hardware and software tasks under the control of a task manager is proposed. The practicality of our system is demonstrated with an FFT application.

Keywords: Codesign, runtime reconfiguration, partitioning and scheduling

Introduction

Reconfigurable hardware has received increasing attention from the research community in the last decade. FPGA-based designs become popular because they provide reconfigurability and short design-time whereas ASIC designs cannot. Instead of using FPGAs simply as ASIC replacements, combining reconfigurable hardware with conventional processors in a codesign system provides an even more flexible and powerful approach for implementing computation intensive applications, and this type of codesign system is the focus of our attention in this paper.

The major concerns in the design process for such codesign systems are synchronization and the integration of hardware and software design [1]. Examples

P. Lysaght and W. Rosenstiel (eds.),
New Algorithms, Architectures and Applications for Reconfigurable Computing, 81–91.

include the partitioning between hardware and software, the scheduling of tasks and the communication between hardware and software tasks in order to achieve the shortest overall runtime. Decisions on partitioning play a major role in the overall system performance and cost. Scheduling is important to ensure completion of all tasks in a real-time environment. These decisions are based heavily on design experience and are difficult to automate. Many design tools leave this burden to the designer by providing semi-auto interactive design environments. A fully automatic design approach for codesign systems is generally considered to be impossible at present. In addition, traditional implementations employ a hardware model that is very different from that used in software. These distinct views of hardware and software tasks can cause problems in the design process. For example, swapping tasks between hardware and software can result in a totally new structure in the control circuit.

This paper presents a new method of constructing and handling tasks in a codesign system. We structure both hardware and software tasks in an interchangeable way without sacrificing the benefit of concurrency found in conventional hardware implementations. The task model in our system exploits the advantages of modularity, scalability, and manageability. We further present a codesign flow containing the partitioning and scheduling algorithms to automate the process of deciding where and when tasks are implemented and run. Our codesign system using this unified hardware/software task model is applied to a reconfigurable computer system known as UltraSONIC [13]. Finally we demonstrate the practicality of our system by implementing an FFT engine.

The contributions of this paper are: 1) a unified way of structuring and modelling hardware and software tasks in a codesign system; 2) a proposed codesign flow for a system with one software and multiple programmable hardware resources; 3) a proposed runtime task manager design to exploit the unified model of tasks; 4) a demonstration of an implementation of a real application using our model in the UltraSONIC reconfigurable platform.

The rest of this paper is organized as follows. Section 7.1 presents some work related to codesign development systems. In section 7.2, we explain how hardware and software tasks are modelled in our codesign system. Section 7.3 describes the codesign flow and the UltraSONIC system used as the realistic target. The system implementation in UltraSONIC and a case study of an FFT algorithm are discussed in section 7.4 and 7.5 respectively. Section 7.6 provides conclusions to this paper.

7.1   Related Work

There are many approaches to hardware/software codesign. Most focus on some particular stages in the design process such as system specification and modelling, partitioning and scheduling, compilation, system co-verification, co-simulation, and code generation for hardware and software interfacing. For example, Ptolemy [2] concentrates on hardware-software co-simulation and system modelling. Chinook [5] focuses on the synthesis of hardware and software interface. MUSIC [6], a multi-language design tool intermediate between SDL and MATLAB, is applied to mechatronic system. For embedded systems, CASTLE [7] and COSYMA [8] design environments are developed.

Codesign system specifically targeted for reconfigurable systems include PAM-Blox [4] and DEFACTO [3]. PAM-Blox is a design environment focusing on hardware synthesis to support PCI Pamette boards that consist of five XC4000 Xilinx FPGAs. This design framework does not provide a complete codesign process. DEFACTO is an end-to-end design environment for developing applications mapped to configurable platforms consisting of FPGAs and a general-purpose microprocessor. DEFACTO concentrates on raising the level of abstraction and developing parallelizing compiler techniques to achieve optimizations on loop transformations and memory accesses. Although both PAM-Blox and DEFACTO are developed specifically for reconfigurable platforms, they take no account of the existence of tasks in the microprocessor.

In this work, tasks can be implemented interchangeably in software or reconfigurable hardware resources. We suggest a novel way for task modelling and task management which will be described later. We also present a novel idea for building infrastructure for dataflow-based applications implementing in a codesign system consisting of a single software and multiple reconfigurable hardware resources. Our approach is inherently modular and is suitable for implementing runtime reconfigurable designs.

7.2   System Architecture

Fig. 7.1 shows the hardware/software system model adopted in this work. We assume the use of a single processor capable of multi-tasking, and a number of concurrent hardware processing elements PE0 to PEn, which are implemented in FPGAs. We employ a *loosely-coupled* model with each processing element (PE) having its own single local memory. All system constraints such as shared resource conflicts, reconfiguration times (of the FPGAs) and communication times are taken into account.

The assumptions used in our model are:

- Tasks implemented in this system are coarse grain tasks which may consist of one or more functional tasks (blocks, loops).

- Because each PE has *one* local memory, only *one* task can access the local memory at any given time. Therefore multiple tasks residing in a given PE must execute sequentially; however, tasks residing across different PEs can execute concurrently.

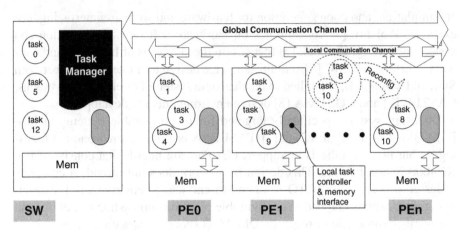

Figure 7.1. Codesign system architecture.

- A temporal group of tasks for a PE can be dynamically swapped through runtime reconfiguration.

- A global communication channel is available for the processor and the PEs to communicate with each other.

- A local communication channel is available for neighboring PEs to communicate with each other.

- There is a well-established mechanism for users to control PEs from the software processor such as a set of API functions.

Because of the reconfigurable capability of the hardware, we can build the hardware tasks very much like software tasks. In this way, the management of task scheduling, task swapping and task allocation can be done in a unified manner, no matter whether the task in question is implemented in hardware or in software. Concurrency is not affected as long as we map concurrent tasks onto separate PEs. Although conceptually different PEs are separate from each other, multiple PEs may be implemented in a single FPGA device.

7.2.1 Task Model

We concentrate on applications dominated by dataflow behavior with few control flow constructs. The applications can be broken down into several coarse grained tasks and are represented as directed acyclic graphs (DAG). Nodes in the graphs represent tasks and edges represent data dependency between tasks. Each task is characterized by its execution time, resource usage, and the cost of its communication with other tasks.

Software and hardware tasks are supposedly encapsulated inside a standard wrapper able to collaborate with the rest of the system. The software tasks are initiated to run in background as worker threads by a *task manager* program. The hardware tasks are assumed to conform to the following restrictions:

- Tasks are non-preemtive. They are executed once according to their precedence levels and priorities.

- Task execution is done in three consecutive steps: *read* input data, *process* the data, and *write* the results. This is done repeatedly until input data stored in memory are all processed. Thus the communication time between memory and an executing task is considered to be a part of the task execution time [9].

- Exactly one task in a given PE is active at any one time. This is a direct consequence of the single port memory restriction. However, multiple tasks may run concurrently provided that they are mapped to different PEs. This is an improvement over the model proposed in [10].

- A task starts executing as soon as all the necessary incoming data from its sources have arrived. It starts writing outgoing data to destinations immediately after processing is complete [11].

7.2.2   Task Manager Program

The center of the control process is the task manager (TM) program running on the processor. This program controls the sequencing of all the tasks, the transfer of data and the synchronization between them, and the dynamic reconfiguration of the FPGA in the PEs when required. Fig. 7.2 shows the conceptual control view of the TM and its operation. The TM communicates with a local task controller on each PE in order to assert control. A message board is used in each PE to receive commands from the TM or to flag completion status to the TM.

In order to properly synchronize the execution of the tasks and the communication between tasks, our task manager employs a *message-passing, event-triggered* protocol when running a process. However, unlike a reactive codesign system [15], we do not use external real-time events as triggers. Instead, we use the termination of each task or data transfer as an event-trigger and signaling of such events is done through dedicated registers. For example, in Fig. 7.2, messages indicating execution completion from tasks 1 are posted to registers inside PE0. The task manager program polls these registers, compiles the messages, and proceeds to the next scheduled operation, in this case running task 3. By using this method, tasks on each PE are run independently because the program operates asynchronously at the system level.

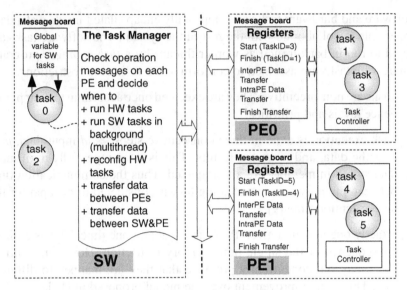

Figure 7.2. The task manager control view.

With a single CPU model, the software processor must simultaneously run the task manager and the software tasks. Multi-threaded programming techniques are then employed to run these two types of processes concurrently.

7.3 Codesign Environment

Fig. 7.3 depicts the codesign environment in our system. The system to be implemented is assumed to be described in some suitable high level language, which is then mapped to a DAG. Tasks are represented by nodes and communications by edges. The nodes are then partitioned into hardware or software tasks and are scheduled to execute in order to obtain the minimum makespan. The partitioning and scheduling software used is adapted from *tabu search* and *list scheduling* algorithms reported earlier [12]. The algorithms are, however, modified to be compatible with this architectural model. A *two-phase clustering* algorithm is used as a pre-processing step to modify the granularity of the tasks in the DAG in order to improve the critical path delay, enable more task parallelism and provide results the achieve 13%–17% shorter makespan [14].

This group of heuristic algorithms, containing clustering, partitioning and scheduling, is collectively called the *CPS* algorithm for short. The CPS algorithm reads textual input files including DAG information and parameter values for clustering, partitioning and scheduling process. During this input stage, the user can optionally specify the type of tasks as *software-only* tasks, *hardware-only* tasks, or *dummy* tasks. A software-only task is a task that the

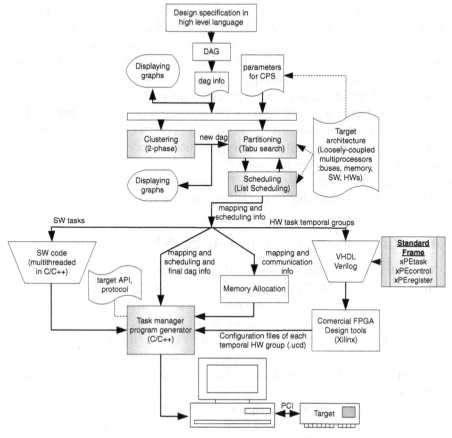

Figure 7.3. Codesign environment.

user intentionally implements in software without exception, and similarly for a hardware-only task. Dummy tasks are either sources or sinks for inputting and outputting data respectively, and are not involved in any computation. In our system, we assume that input and output data are initially provided and written to the host processor memory. Unspecified tasks are then free to be partitioned, scheduled and bound to either hardware or software resources.

The result of the partitioning and scheduling processes are the physical and temporal bindings for each task. Note that in the case of hardware tasks, they may be divided into many temporal groups that can either be statically mapped to the hardware resource, or dynamically configured during runtime.

We currently assume that software tasks are manually written in C/C++, while hardware tasks are designed manually in a HDL (such as Verilog in this work) using a library-based approach. Once all the hardware tasks for a given PE are available, they are implemented in a standard frame which is a pre-designed circuit consisting of xPEtask, xPEregister and xPEcontrol, and is application-independent. Commercially available synthesis and place-and-route tools are then used to produce the final configuration files for each hardware module. Each task in this implementation method requires some hardware overhead to implement the task frame wrapper circuit. Therefore our system favours partitioning algorithms that generate coarse grained tasks.

The results from the partitioning and scheduling process, the memory allocator, the task control protocol, the API functions, and the configuration files of hardware tasks are used to automatically generate the code for the task manager program that controls all operations in this system, such as dynamic configuration, task execution and data transfer. The resulting task manager is inherently multi-threaded to ensure that tasks can run concurrently.

7.4 Implementation in the UltraSONIC Platform

The codesign methodology described above is targeted for the UltraSONIC System [13]. UltraSONIC (see Fig. 7.4(a)) is a reconfigurable computing system designed to cope with the computational power and the high data throughput demanded by real-time video applications. The system consists of Plug-In Processing Elements (PIPEs) interconnected by local and global buses. The architecture exploits the spatial and the temporal parallelism in video processing algorithms. It also facilitates design reuse and supports the software plug-in methodology.

Fig. 7.4(b) shows how our codesign model is implemented in the UltraSONIC PIPE. xPEcontrol is used to control local operations such as task execution and local communication with other PEs. To run a task, for example, all necessary

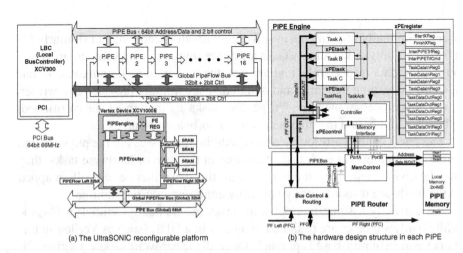

(a) The UltraSONIC reconfigurable platform   (b) The hardware design structure in each PIPE

Figure 7.4.   The hardware implementation in the UltraSONIC architecture.

information for the task execution is first sent to xPEregister (used as a message board) by the task manager. After that, the task can be triggered by sending a unique identification of the task (called TaskID) to the *StartXReg* register. It is worth noting that all the hardware tasks in the PIPE engine are wrapped inside xPEtask. When xPEtask, which is designed to monitor this register all the time, discovers that the TaskID of the task inside is the same as the register value, the task execution process is initiated. The task communicates with xPEcontrol to interface with the local memory. When execution is finished, the task notifies the task manager program in order to continue the next process by writing the TaskID to the *FinishXReg* register.

The total hardware overhead of this infrastructure is modest. xPEcontrol and xPEregister consume around 6% and 4% of the reconfigurable resource on each PIPE (Xilinx's XCV1000E) respectively. The hardware overhead of xPEtask depends on the size of the I/O buffers required for a task.

7.5   A Case Study of FFT Algorithm

In order to demonstrate the practicality of our system, we chose to implement the well-known FFT algorithm. Although we can implement the algorithm for an arbitrary data length, we use an 8-point FFT implementation to illustrate the results of our codesign system. The DAG of an 8-point FFT can be extracted as shown in Fig. 7.5(a). Nodes 0, 1, 2 are used for arranging input data and are implemented as software-only tasks. Tasks 3 to 14 are butterfly computation nodes which can be implemented in either hardware or software. The number shown inside the parenthesis (the second number) on each edge is the amount of data needed for each execution of a task. Each task is executed repeatedly until

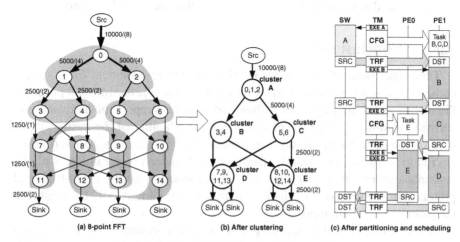

Figure 7.5.   The DAG of 8-point FFT algorithm.

all data (shown as the first number on the edge) are processed. Initially in this DAG, the software execution times are obtained by profiling tasks on the PC, while the hardware times and areas are obtained by using Xilinx development tools.

This DAG information is supplied to the CPS algorithm in our design environment (see Fig. 7.3) to perform automatic clustering, partitioning and scheduling. Parameters such as reconfiguration time, bus speed, and FPGA size are all based on the UltraSONIC system. The clustering algorithm produces a new DAG that contains tasks in higher granularity as shown in Fig. 7.5(b). These new tasks are then iteratively partitioned and scheduled. The computational part of each of the hardware tasks is designed manually in Verilog and the software tasks are written in C. Our tools then combine the results from the partitioning and scheduling algorithms to automatically generate the task manager program.

Fig. 7.5(c) depicts an example of the scheduling profile of this implementation. Each column represents activities on the available resources which are software tasks (SW), and two hardware processing elements (PE0 and PE1). TM is the task manager program which is also running on the software resource. The execution of this algorithm proceeds from top to bottom. It shows all the runtime activities including configuration (CFG), transfer of data (TRF), events that trigger task executions (EXE), the supply of data (SRC), data reception (DST) and the executions of the tasks (A to E).

The FFT algorithm for different data window sizes are also tested on the UltraSONIC system and are shown to work correctly. The method, although it requires some manual design steps, is very quick. Implementing the FFT algorithm only took a few hours from specification to completion. We also successfully implemented the standard baseline method of JPEG compression (CCITT Recommendation T.81) for colour images by using this codesign environment. Nonetheless, due to the lack of space, the details cannot be described here.

7.6   Conclusions

This paper presents a semi-automatic codesign environment for a system consisting of single software and multiple reconfigurable hardware. It proposes the use of a task manager to combine the runtime support for hardware and software in order to improve modularity and scalability of the design. Partitioning and scheduling are done automatically. Programs for software tasks are run in software concurrently (using multi-threaded programming) with the task manager program which is based on a message-passing, event-triggered protocol. Implementation of the FFT algorithm on UltraSONIC demonstrates the practicality of our approach.

Future work includes applying our codesign system with more complex applications, mapping behavioral or structural descriptions to DAGs automatically, improving the task management environment so that task pipelining can also be handled, and migrating our codesign system to System-on-a-Programmable-Chip (SoPC).

Acknowledgments

The authors would like to acknowledge the continuing support of John Stone, Simon Haynes, Henry Epsom and the rest of the UltraSONIC team at Sony Broadcast Professional Research Laboratory in UK.

References

[1] R. Ernst "Codesign of embedded systems: status and trends", *IEEE Design & Test of Computers*, 1998.

[2] G. Manikutty, and H. Hanson, "Hardware/Software Partitioning of Synchronous Dataflow Graphs in the ACS domain of Ptolemy", University of Texas, Literature Survey, Final Report May 12 1999.

[3] M. Hall, P. Diniz, K. Bondalapati, H. Ziegler et al., "DEFACTO:A Design Environment for Adaptive Computing Technology", *Proceedings of the 6th Reconfigurable Architectures Workshop*, 1999.

[4] O. Mencer, M. Morf, and M. J. Flynn, "PAM-Blox: high performance FPGA design for adaptive computing", *FPGAs for Custom Computing Machines*, 1998.

[5] P. H. Chou, R. B. Ortega, and G. Borriello, "The Chinook hardware/software co-synthesis system", *System Synthesis*, 1995.

[6] P. Coste, F. Hessel, P. Le Marrec, Z. Sugar et al., "Multilanguage design of heterogeneous systems", *Hardware/Software Codesign*, 1999.

[7] J. Wilberg, A. Kuth, R. Camposano, W. Rosenstiel et al., "Design Exploration in CASTLE", *Workshop on High Level Synthesis Algorithms Tools and Design (HILES)*, 1995.

[8] R. Ernst, "Hardware/Software Co-Design of Embedded Systems", *Asia Pacific Conference on Computer Hardware Description Languages*, 1997.

[9] T. Pop, P. Eles, and Z. Peng, "Holistic scheduling and analysis of mixed time/event-triggered distributed embedded systems", *Hardware/Softwarw Codesign*, 2002.

[10] V. Srinivasan, S. Govindarajan, and R. Vemuri, "Fine-grained and coarse-grained behavioral partitioning with effective utilization of memory and design space exploration for multi-FPGA architectures", *IEEE Transactions on Very Large Scale Integration (VLSI) Systems*, vol. 9, pp. 140–158, 2001.

[11] J. Hou, and W. Wolf, "Process partitioning for distributed embedded systems", *Hardware/Software Co-Design*, 1996.

[12] T. Wiangtong, P. Y. K. Cheung, and W. Luk, "Comparing Three Heuristic Search Methods for Functional Partitioning in HW-SW Codesign", *International Journal on Design Automation for Embedded Systems*, vol. 6, pp. 425–449, July 2002.

[13] S. D. Haynes et al., "UltraSONIC: A Reconfigurable Architecture for Video Image Processing", *Field-Programmable Logic and Applications (FPL)*, 2002.

[14] T. Wiangtong, P. Y. K. Cheung, and W. Luk, "Cluster-Driven Hardware/Software Partitioning and Scheduling Approach For a Reconfigurable Computer System", *Field-Programmable Logic and Applications (FPL)*, 2003.

[15] G. D. Micheli, "Computer-aided hardware-software codesign", *IEEE Micro*, Vol 14, pp. 10–16, 1994.

Chapter 8

Mapping Applications to a Coarse Grain Reconfigurable System

Yuanqing Guo, Gerard J.M. Smit, Michèl A.J. Rosien, Paul M. Heysters, Thijs Krol, Hajo Broersma

University of Twente, Faculty of Electrical Engineering,
Mathematics and Computer Science, P.O. Box 217, 7500AE Enschede,
The Netherlands
{yguo, smit, rosien, heysters, krol}@cs.utwente.nl; h.j.broersma@utwente.nl

Abstract   This paper introduces a design method to map applications written in a high level source language program, like C, to a coarse grain reconfigurable architecture, called MONTIUM. The source code is first translated into a control dataflow graph (CDFG). Then after applying graph clustering, scheduling and allocation on this CDFG, it can be mapped onto the target architecture. High performance and low power consumption are achieved by exploiting maximum parallelism and locality of reference respectively. Using our mapping method, the flexibility of the MONTIUM architecture can be exploited.

Keywords: Coarse Grain Reconfigurable Architecture, Mapping and Scheduling

8.1   Introduction

In the CHAMELEON/GECKO project we are designing a heterogeneous reconfigurable System-On-Chip (SoC) [Smit, 2002] for 3G/4G terminals. This SoC contains a general-purpose processor (ARM core), a bit-level reconfigurable part (FPGA) and several word-level reconfigurable parts (MONTIUM tiles). We believe that in future 3G/4G terminals heterogeneous reconfigurable architectures are needed. The main reason is that the efficiency (in terms of performance or energy) of the system can be improved significantly by mapping application tasks (or kernels) onto the most suitable processing entity.

The objective of this paper is to present a design method for mapping processes, written in a high level language, to a reconfigurable platform. The

P. Lysaght and W. Rosenstiel (eds.),
New Algorithms, Architectures and Applications for Reconfigurable Computing, 93–103.
© 2005 *Springer. Printed in the Netherlands.*

method can be used to optimize the system with respect to certain criteria e.g. energy efficiency or execution speed.

8.2 The Target Architecture: MONTIUM

In this section we give a brief overview of the MONTIUM architecture, because this architecture led to the research questions and the algorithms presented in this paper. Figure 8.1 depicts a single MONTIUM processor tile. The hardware organization within a tile is very regular and resembles a very long instruction word (VLIW) architecture. The five identical arithmetic and logic units (ALU1 ⋯ ALU5) in a tile can exploit spatial concurrency to enhance performance. This parallelism demands a very high memory bandwidth, which is obtained by having 10 local memories (M01 ⋯ M10) in parallel. The small local memories are also motivated by the locality of reference principle. The ALU input registers provide an even more local level of storage. Locality of reference is one of the guiding principles applied to obtain energy-efficiency in the MONTIUM. The MONTIUM has a datapath width of 16-bits and supports both integer and fixed-point arithmetic. Each local SRAM is 16-bit wide and has a depth of 512 positions, which adds up to a storage capacity of 8 Kbit per local memory. A memory has only a single address port that is used for both reading and writing. A reconfigurable address generation unit (AGU) accompanies each memory. The AGU contains an address register that can be modified using base and modify registers.

A single ALU has four 16-bit inputs. Each input has a private input register file that can store up to four operands. The input register file cannot be by-passed, i.e., an operand is always read from an input register. Input registers can be written by various sources via a flexible interconnect. An ALU has two 16-bit outputs, which are connected to the interconnect. The ALU is entirely combinatorial and consequentially there are no pipeline registers within the ALU. Each MONTIUM ALU contains two different levels. Level 1 contains four

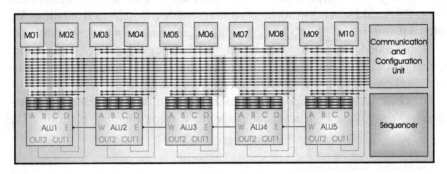

Figure 8.1. MONTIUM processor tile.

function units. A function unit implements the general arithmetic and logic operations that are available in languages like C (except multiplication and division). Level 2 contains the MAC unit and is optimised for algorithms such as FFT and FIR. Levels can be bypassed (in software) when they are not needed.

Neighboring ALUs can also communicate directly on level 2. The West-output of an ALU connects to the East-input of the ALU neighboring on the left (the West-output of the leftmost ALU is not connected and the East-input of the rightmost ALU is always zero). The 32-bit wide East-West connection makes it possible to accumulate the MAC result of the right neighbor to the multiplier result (note that this is also a MAC operation). This is particularly useful for performing a complex multiplication, or for adding up a large amount of numbers (up to 20 in one clock cycle). The East-West connection does not introduce pipeline, as it is not registered.

8.3   A Four-Phase Decomposition

The overall aim of our research is to execute DSP programs written in high level language, such as C, by one MONTIUM tile in as few clock cycles as possible. There are many related aspects: the limitation of resources; the size of total configuration space; the ALU structure etc. The main question is to find an optimal solution under all those constraints. Therefore we propose to decompose this problem into a number of phases: Based on the two-phased decomposition of multiprocessor scheduling introduced by [Sarkar, 1989], our work is built on a four-phase decomposition: translation, clustering, scheduling and resource allocation:

1 **Translating the source code to a CDFG**: The input C program is first translated into a CDFG; and then some transformations and simplifications are done on the CDFG. The focus of this phase is the input program and is largely independent of the target architecture.

2 **Task clustering and ALU data-path mapping**, clustering for short: The CDFG is partitioned into clusters and mapped to an unbounded number of fully connected ALUs. The ALU structure is the main concern of this phase and we do not take the inter-ALU communication into consideration;

3 **Scheduling**: The graph obtained from the clustering phase is scheduled taking the maximum number of ALUs (it is 5 in our case) into account. The algorithm tries to find the minimize number of the distinct configurations of ALUs of a tile;

4 **Resource allocation**, allocation for short: The scheduled graph is mapped to the resources where locality of reference is exploited, which is

important for performance and energy reasons. The main challenge in this phase is the limitation of the size of register banks and memories, the number of buses of the crossbar and the number of reading and writing ports of memories and register banks.

Note that when one phase does not give a solution, we have to fall back to a previous phase and select another solution.

8.4 Translating C to a CDFG

A control data flow graph (CDFG) $G = (N_G, P_G, A_G)$ consists of two finite non-empty sets of **nodes** N_G and **ports** P_G and a set A_G of so-called **hydra-arcs**; a hydra-arc $a = (t_a, H_a)$ has one **tail** $t_a \in N_G \cup P_G$ and a non-empty set of **heads** $H_a \subset N_G \cup P_G$. In our applications, N_G represents the operations of a CDFG, P_G represents the inputs and outputs of the CDFG, while the hydra-arc (t_a, H_a) either reflects that an input is used by an operation (if $t_a \in P_G$), or that an output of the operation represented by $t_a \in N_G$ is input of the operations represented by H_a, or that this output is just an output of the CDFG (if H_a contains a port of P_G).

See the example in Figure 8.2: The operation of each node is a basic computation such as addition (in this case), multiplication, or subtraction. Hydra-arcs are directed from their tail to their heads. Because an operand might be input for more than one operation, a hydra-arc is allowed to have multiple heads although it always has only one tail. The hydra-arc $e7$ in Figure 8.2, for instance, has two heads, w and v. The CDFG communicates with external systems through its ports represented by small grey circles in Figure 8.2.

In general, CDFGs are not acyclic. In the first phase we decompose the general CDFG into acyclic blocks and cyclic control information. The control

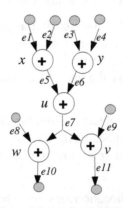

Figure 8.2. A small CDFG.

```
const int n = 4;
float x_re[n] ; float x_im[n] ; float w_re[n/2]; float w_im[n/2];
void main() {
  int xi, xip;
  float u_re, u_im, x_re_tmp_xi, x_re_tmp_xip, x_im_tmp_xi, x_im_tmp_xip;
  for (int le = n / 2; le > 0; le /= 2) {
    for (int j = 0; j < le; j++) {
      int step = n / le;
      for (int i = 0; i < step/2; i++ ) {
        xi = i + j * step; xip = xi + step/2; u_re = w_re[le * i]; u_im = w_im[le * i];
        x_re_tmp_xi = x_re[xi]; x_re_tmp_xip = x_re[xip];
        x_im_tmp_xi = x_im[xi]; x_im_tmp_xip = x_im[xip];
        x_re[xi] = x_re_tmp_xi + (u_re * x_re_tmp_xip - u_im * x_im_tmp_xip);
        x_re[xip] = x_re_tmp_xi - (u_re *x_re_tmp_xip - u_im * x_im_tmp_xip);
        x_im[xi] = x_im_tmp_xi + (u_re * x_im_tmp_xip + u_im * x_re_tmp_xip);
        x_im[xip] = x_im_tmp_xi  - (u_re * x_im_tmp_xip+ u_im * x_re_tmp_xip);
      }
    }
  }
}
```

Figure 8.3.   C code for the *n*-point FFT algorithm.

information part of the CDFG will be handled by the sequencer of the MONTIUM. In this paper we only consider acyclic parts of CDFGs. To illustrate our approach, we use an FFT algorithm. The Fourier transform algorithm transforms a signal from the time domain to the frequency domain. For digital signal processing, we are particularly interested in the discrete Fourier transform. The fast Fourier transform (FFT) can be used to calculate a DFT efficiently. The source C code of a *n*-point FFT algorithm is given in Figure 8.3 and Figure 8.4 shows the CDFG generated automatically from a piece of 4-point FFT code after C code translation, simplification and complete loop expansion. This example will be used throughout this paper.

8.5   Clustering

In the clustering phase the CDFG is partitioned and mapped to an unbounded number of fully connected ALUs, i.e., the inter-ALU communication is not yet considered. A cluster corresponds to a possible configuration of an ALU data-path, which is called **one-ALU configuration**. Each one-ALU configuration has fixed input and output ports, fixed function blocks and fixed control signals. A partition with one or more clusters that can not be mapped to our MONTIUM ALU data-path is a failed partition. For this reason the procedure of clustering should be combined with ALU data-path mapping. Goals of clustering are 1) minimization of the number of ALUs required; 2) minimization of the number of distinct ALU configurations; and 3) minimization of the length of the critical path of the dataflow graph.

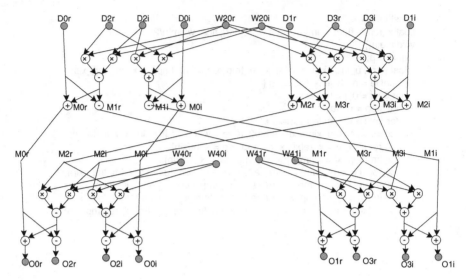

Figure 8.4. The generated CDFG of a 4-point FFT after complete loop unrolling and full simplification.

The clustering phase is implemented by a graph-covering algorithm [Guo, 2003]. The procedure of clustering is the procedure of finding a cover for a CDFG which is implemented in two steps:

Step A: Template Generation Problem

Given a CDFG, generate the complete set of nonisomorphic templates (that satisfy certain properties, e.g., which can be executed on the ALU-architecture in one clock cycle), and find all their corresponding matches (instances).

Step B: Template Selection Problem

Given a CDFG G and a set of (matches of) templates, find an 'optimal' cover of G.

See [Guo, 2003] for the details of the graph-covering algorithm.

Each selected match is a cluster that can be mapped onto one MONTIUM ALU and can be executed in one clock-cycle [Rosien, 2003]. As an example Figure 8.5 presents the produced cover for the 4-point FFT. The letters inside the dark circles indicate the templates. The graph is completely covered by three templates. This result is the same as our manual solution. It appears that the same templates are chosen for a n-point FFT ($n = 2^d$).

8.6 Scheduling

To facilitate the scheduling of clusters, all clusters get a **level** number. The level numbers are assigned to clusters with the following restrictions:

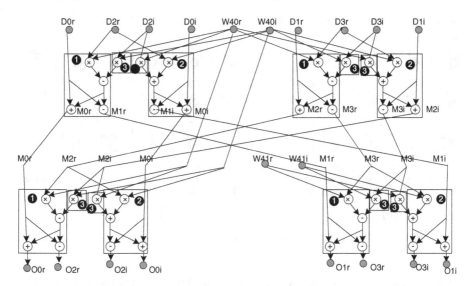

Figure 8.5. The selected cover for the CDFG in 8.4.

- For a cluster A that is dependent on a cluster B with level number i, cluster A must get a level number $> i$ if the two clusters cannot be connected by the west-east connection.

- Clusters that can be executed in parallel may have equal level numbers.

- Clusters that depend only on in-ports have level number one.

The objective of the clustering phase is to minimize the number of different configurations for separate ALUs, i.e. to minimize the number of different one-ALU configurations. The configurations for all five ALUs of one clock cycle form a **5-ALU configuration**. Since our MONTIUM tile is a very long instruction word (VLIW) processor, the number of distinct 5-ALU configurations should be minimized as well. At the same time, the maximum amount of parallelism is preferable within the restrictions of the target architecture. In our architecture, at most 5 clusters can be on the same level.

If there are more than 5 clusters at some level, one or more clusters should be moved one level down. Sometimes one or more extra clock cycles have to be inserted. Take Figure 8.5 as an example, where, in level one, the clusters of type 1 and type 2 are dependent on clusters of type 3. However, by using the east-west connection, clusters of type 1/2 and type 3 can be executed on the same level. Because there are too many clusters in level 1 and level 2 of Figure 8.5, we have to split them. Figure 8.6(a) shows a possible scheduling scheme where not all five ALUs are used. This scheme consists of only one 5-ALU

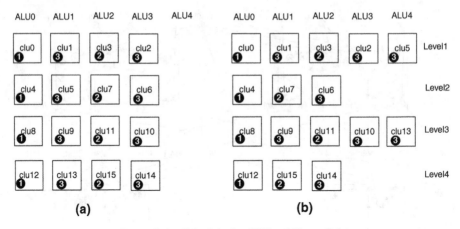

Figure 8.6. Schedule the ALUs of Figure 8.5.

configuration: $C1 = \{$ ❶❸❷❸○ $\}$. As a result, with the scheme of 8.6(a), the configuration of ALUs stays the same during the execution. The scheduling scheme of Figure 8.6(b) consists of 4 levels as well, but it is not preferable because it needs two distinct 5-ALU configurations: $C2 = \{$ ❶❸❷❸❸ $\}$ and $C3 = \{$ ❶❷❸○○ $\}$. Switching configurations adds to the energy and control overhead.

8.7 Allocation

The main architectural issues of the MONTIUM that are relevant for the resource allocation phase are summarized as follows:

(1) The size of a memory is 512 words; (2) Each register bank includes 4 registers; (3) Only one word can be read from or written to a memory within one clock cycle; (4) The crossbar has a limited number of buses (10); (5) The execution time of the data-path is fixed (one clock cycle); (6) An ALU can only use the data from its local registers or from the east connection as inputs.

After scheduling, each cluster is assigned an ALU and the relative executing order of clusters has been determined. In the allocation phase, the other resources (busses, registers, memories, etc) are assigned, where locality of reference is exploited, which is important for performance and energy reasons. The main challenge in this phase is the limitation of the size of register banks and memories, the number of buses of the crossbar and the number of reading and writing ports of memories and register banks. The decisions that should be made during allocation phase are:

- Choose proper storage places (memories or registers) for each intermediate value;

- Arrange the resources (crossbar, address generators, etc) such that the outputs of the ALUs are stored in the right registers and memories;

- Arrange the resources such that the inputs of ALUs are in the proper register for the next cluster that will execute on that ALU.

Storing an ALU result must be done in the clock cycle within which the output is computed. When the outputs are not moved to registers or memories immediately after generated by ALUs, they will be lost. For this reason, in each clock cycle, storing outputs of the current clock cycle takes priority over using the resources. Preparing an input should be done at least one clock cycle before it is used. However, when it is prepared too early, the input will occupy the register space for a too long time. A proper heuristic is starting to prepare an input 4 clock cycles before the clock cycle it is actually used by the ALU. If the inputs are not well prepared before the execution of an ALU, one or more extra clock cycles need to be inserted to do so. However, this will decrease the runtime of the algorithm.

When a value is moved from a memory to a register, a check should be done whether it is necessary to keep the old copy in the memory or not. In most cases, a memory location can be released after the datum is fed into an ALU. An exception is when there is another cluster which shares the copy of the datum and that cluster has not been executed.

We adopt a heuristic resource allocation method, whose pseudocode is listed in Figure 8.7. The clusters in the scheduled graph are allocated level by level (lines 0–2). Firstly, for each level, the ALUs are allocated (line 4). Secondly, the outputs are stored through the crossbar (line 5). As said before, storing outputs is given highest priority. The locality of reference principle is employed again to choose a proper storage position (register or memory) for each output. The

```
//Input: Scheduled Clustered Graph G
//Output: The job of an FPFA tile for each clock cycle
0   function ResourseAllocation(G) {
1       for each level in G do  Allocate(level);
2   }
3   function Allocate(currentLevel) {
4       Allocate ALUs of the current clock cycle
5       for each output  do store it to a memory;
6       for each input of current level
7       do try to move it to proper register at the clock cycle which is four steps
8           before; If failed,  do it three steps before; then two steps before;  one
9           step before.
10      if some inputs are not moved successfully
11      then insert one or more clock cycles before the current one to load inputs
12  }
```

Figure 8.7. Pseudocode of the heuristic allocation algorithm.

Table 8.1. The resource allocation result for the 4-point FFT CDFG

Step	Actions
1	Load inputs for clusters of level 1 in Figure 8.6.
2	Clu0, Clu1, Clu2 and Clu3 are executed; Save outputs of step 2; Load inputs for clusters of level 2.
3	Clu4, Clu5, Clu6 and Clu7 are executed; Save outputs of step 3; Load inputs for clusters of level 3.
4	Clu8, Clu9, Clu10 and Clu11 are executed; Save outputs of step 4; Load inputs for clusters of level 4.
5	Clu12, Clu13, Clu14 and Clu15 are executed; Save outputs of step 5.

unused resources (busses, registers, memories) of previous steps are used to load the missing inputs (lines 6–9) for the current step. Finally, extra clock cycles might be inserted if some inputs could not be put in place by the preceding steps (lines 10–11).

The resource allocation result for the 4-point FFT CDFG is listed in Table 8.1. Before the execution of Clu0, Clu1, Clu2 and Clu3, an extra step (step 1) is needed to load their inputs to proper local registers. In all other steps, besides saving the result of current step, the resources are sufficient to loading the inputs for the next step, so no extra steps are needed. The 4-point FFT can be executed within 5 steps by one MONTIUM tile. Note that when a previous algorithm already left the input data in the right registers, step 1 is not needed and consequently the algorithm can be executed in 4 clock cycles.

8.8 Conclusion

In this paper we presented a method to map a process written in a high level language, such as C, to one MONTIUM tile. The mapping procedure is divided into four steps: translating the source code to a CDFG, clustering, scheduling and resource allocation. High performance and low power consumption are achieved by exploiting maximum parallelism and locality of reference respectively. In conclusion, using this mapping scheme the flexibility and efficiency of the MONTIUM architecture are exploited.

8.9 Related work

High level compilation for reconfigurable architectures has been the focus of many researchers, see [Callahan, 2000]. Most systems use the SUIF compiler of Stanford [http://suif.stanford.edu].

The procedure of mapping a task graph to a MONTIUM tile has NP complexity just like the task scheduling problem on multiprocessors systems [Rewini, 1994]. [Kim, 1988] considered linear clustering which is an important special case of clustering. [Sarkar, 1989] presents a clustering algorithm based on a scheduling algorithm on unbounded number of processors. In [Yang, 1994] a fast and accurate heuristic algorithm was proposed, the Dominant Sequence Clustering. The ALUs within our MONTIUM are interconnected more tightly than multiprocessors. This difference prevents us from using their solution directly.

To simplify the problem, we use a four-phased decomposition algorithm based on the two-phased decomposition of multiprocessor scheduling introduced by [Sarkar, 1989].

Acknowledgments

This research is conducted within the Chameleon project (TES.5004) and Gecko project (612.064.103) supported by the PROGram for Research on Embedded Systems & Software (PROGRESS) of the Dutch organization for Scientific Research NWO, the Dutch Ministry of Economic Affairs and the technology foundation STW.

References

T.J. Callahan, J.R. Hauser, and J. Wawryzynek, "The Garp Architecture and C compiler" in *IEEE Computer*, 33(4), April 2000.

Yuanqing Guo, Gerard Smit, Paul Heysters, Hajo Broersma, "A Graph Covering Algorithm for a Coarse Grain Reconfigurable System", *2003 ACM Sigplan Conference on Languages, Compilers, and Tools for Embedded Systems(LCTES'03)*, California, USA, June 2003, pp. 199–208.

S.J. Kim and J.C. Browne, A General Approach to Mapping of parallel Computation upon Multiprocessor Architectures, *International Conference on Parallel Processing,* vol 3, 1988, pp. 1–8.

Hesham EL-Rewini, Theodore Gyle Lewis, Hesham H. Ali, *Task scheduling in parallel and distributed systems,* PTR Prentice Hall, 1994.

Michel A.J. Rosien, Yuanqing Guo, Gerard J.M. Smit, Thijs Krol, "Mapping Applications to an FPFA Tile", *Proc. of Date 03*, Munich, March, 2003.

Vivek Sarkar. *Clustering and Scheduling Parallel Programs for Multiprocessors*. Research Monographs in Parallel and Distributed Computing. MIT Press, Cambridge, Massachusetts, 1989.

Gerard J.M. Smit, Paul J.M. Havinga, Lodewijk T. Smit, Paul M. Heysters, Michel A.J. Rosien, "Dynamic Reconfiguration in Mobile Systems", *Proc. of FPL2002*, Montpellier France, pp. 171–181, September 2002.

SUIF Compiler system, http://suif.stanford.edu.

Tao Yang; Apostolos Gerasoulis, "DSC: scheduling parallel tasks on an unbounded number of processors", *IEEE Transactions on Parallel and Distributed Systems*, Volume:5 Issue:9, Sept. 1994 Page(s): 951–967.

The procedure of mapping task graph to a MONTIUM tile on NET core processor pair on the PROCESS core, using available in the multiprocessor system. It was (1994; Kian, 1998) considered linear clustering which is an important special case of clustering (Kim et al. 1988) presents a clustering algorithm based on a clustering algorithm on unbounded number of processors. In (Kung, 1988) a list and greedy clustering algorithm was proposed. The Dominant Sequence Clustering algorithm from our MONTIUM architecture used more than that performance. The difference or value is from a structure resolution of the future.

To simplify the problem, we use a load-based decomposition algorithm based on the two-phased deterministic list scheduling for scheduling routine directly (Dick, 1998).

Acknowledgements

This research conducted with the CardioLab project (TES.6021) and by the Embedded Systems Support by the PROGram for Research in Embedded Systems & Software (PROGRESS) of the Dutch Technology Foundation STW, the Dutch Ministry of Economic Affairs, and the Dutch Applied Research Foundation (STW).

References

[Illegible reference entries]

Chapter 9

Compilation and Temporal Partitioning for a Coarse-grain Reconfigurable Architecture

João M.P. Cardoso[1], Markus Weinhardt[2]

[1] *Faculty of Sciences and Technology, University of Algarve, Campus de Gambelas 8000—117 Faro, Portugal*
jmpc@acm.org

[2] *PACT XPP Technologies AG, Muthmannstr. 1, 80939 München, Germany*
m.weinhardt@computer.org

Abstract   The eXtreme Processing Platform (XPP) is a coarse-grained dynamically reconfigurable architecture. Its advanced reconfiguration features make feasible the configure-execute paradigm, the natural paradigm of dynamically reconfigurable computing. This chapter presents a compiler aiming to program the XPP using a subset of the C language. The compiler, apart from mapping the computational structures onto the available resources on the device, splits the program in temporal sections when it needs more resources than the physically available. In addition, since the execution of the computational structures in a configuration needs at least two stages (e.g., configuring and computing), a scheme to split the program such that the reconfiguration overheads are minimized, taking advantage of the overlapping of the execution stages on different configurations is presented.

Keywords: Compilation, reconfigurable computing, temporal partitioning

9.1   Introduction

Run-time reconfigurable architectures promise to be efficient solutions when flexibility, high-performance, and low power consumption are required features. As Field-Programmable Gate Arrays (FPGAs) require long design cycles, with low level hardware efforts, and the fine-granularity (bit level) used does not match many of today's applications, new reconfigurable processing units are being introduced [1]. The XPP (eXtreme Processing Platform), a coarse-grained dynamically reconfigurable architecture, is one of those units [2]. However,

P. Lysaght and W. Rosenstiel (eds.),
New Algorithms, Architectures and Applications for Reconfigurable Computing, 105–115.

compilation research efforts are still required in order to facilitate programming of those architectures. Compilers must both reduce the time to program such units, hide the architecture's details from the user, and efficiently map algorithms to the available resources.

This chapter presents techniques to compile programs in a C subset to the XPP [3], [4]. The techniques include forms to efficiently map algorithms to the XPP structures (e.g., vectorization techniques [5]) and the use of temporal partitioning schemes for the efficient generation of multiple configurations. Temporal partitioning allows to map large programs. Besides, since the execution of the computational structures in a configuration needs at least two stages (configuring and computing), the temporal partitioning techniques try to split the program such that the reconfiguration overheads are minimized, taking advantage of the overlapping of the execution stages of different configurations.

This chapter is organized as follows. The next section briefly introduces and explains the XPP architecture and the configure-execute paradigm. Section 3 describes the compilation flow, which includes the mapping of computational structures to a single configuration, and the generation of multiple configurations. Section 4 shows some experimental results, and section 5 describes the related work. Finally, section 6 concludes the chapter.

9.2 The XPP Architecture and the Configure-Execute Paradigm

The XPP architecture (see Fig. 9.1) is based on a 2D array of coarse-grained adaptive Processing Array Elements (PAEs), I/O ports, and interconnection resources [2]. A PAE contains three elements: one FREG (forward register), one BREG (backward register), and one ALU or memory (MEM). The FREG is used for vertical and registered forward routing, or to perform MERGE, SWAP or DEMUX operations (for controlled stream manipulations). The BREG is used for vertical backward routing (registered or not), or to perform some selected arithmetic operations (e.g., ADD, SUB, SHIFT). The BREGs can also be used to perform logical operations on events, i.e. one-bit control signals. Each ALU performs common two-input arithmetic and logical operations, and comparisons. A MAC (multiply and accumulate) operation can be performed using the ALU and the BREG of one PAE in a single clock cycle. In each configuration, each ALU performs one dedicated operation.

The interconnection resources include several segmented busses which can be configured to connect the output of a PAE with other PAEs' inputs. Separate event and data busses are used to communicate 1-bit events and n-bit data values, respectively. The operator bit-width n is implementation dependent.

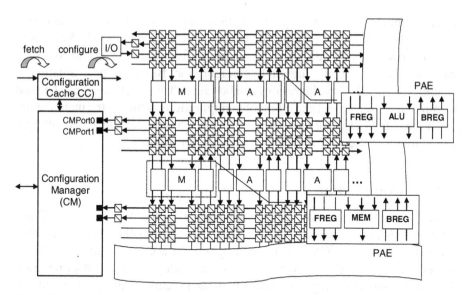

Figure 9.1.   XPP architecture. Note that this abstract view hides many architecture details.

The interface of the array with other units uses the I/O ports. They can be used to directly stream data to/from external channels or memories. The XPP is a data-driven architecture where the flow of data is synchronized by a handshake protocol.

The array is coupled with a Configuration Manager (CM) responsible for the run-time management of configurations and for loading configuration data into the configurable resources of the array. The CM has cache memory for accommodating multiple configurations.

The XPP has some similarities with other coarse-grained reconfigurable architectures (e.g., KressArray [6]). XPP's main distinguishing features are its sophisticated self-synchronization and configuration mechanisms. Although it is a data-driven architecture, the XPP permits to map computations with complex imperative control flow (e.g., nested loops) and with irregular data accesses.

The execution of a configuration in the XPP requires three stages: fetch, configure, and compute. They are related to loading the configuration data to the CM cache, programming the array structures, and execution on the array, respectively. The compute stage includes the release of the used resources after completion of execution. The self-release of the resources enables the execution of an application consisting of several configurations without any external control. When a computation is finished, the PAE detecting the termination (e.g., a counter) can generate an event to be broadcast or can send an event to the CM requesting the next configuration. The stages can be executed in parallel for different configurations (e.g., during the configure or compute stages of one configuration the fetch of another configuration can be performed). Since only the array resources actually used have to be configured, the fetch and configure latencies vary from configuration to configuration. Based on the evaluation of conditions, different configurations can be called. This mechanism can be used to implement conditional execution of configurations, which is the main feature to exploit if-then-else stuctures where each branch is implemented in a different configuration (used in "loop dissevering" [3], for instance).

9.3   Compilation

The overall compilation flow, implemented in the Vectorizing C Compiler, XPP-VC, is shown in Fig. 9.2. The compiler uses the SUIF framework [7] as the front-end. XPP-VC maps programs in a C subset to the XPP resources. The currently supported C subset excludes struct and floating-point data types, pointers, irregular control flow, and recursive and system calls. A compiler options file specifies the parameters of the target XPP and the external memories connected to it. To access XPP I/O ports specific C functions are provided.

Compilation is performed using a number of steps. First, architecture-independent preprocessing passes based on well-known compilation techniques [8] are performed (e.g., loop unrolling). Then, the compiler performs a data-dependence analysis and tries to vectorize inner FOR loops. In XPP-VC, vectorization means that loop iterations are overlapped and executed in a pipelined, parallel fashion. The technique is based on the Pipeline Vectorization method [5]. Next, TempPart splits the C program into several temporal partitions if required. The MODGen task of the compiler transforms a sequential C description to an efficient data- and event-driven control/dataflow graph (CDFG) which can be directly mapped to the structures of the XPP array. MODGen generates one NML (the native language of the XPP) module for each partition. Each NML module is placed and routed automatically by PACT's proprietary xmap tool, which uses an enhanced force-directed placer with short runtimes.

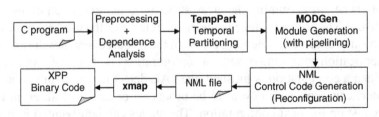

Figure 9.2.   XPP-VC compilation flow.

Mapping C Code to XPP Resources

To transform C code to NML code a number of steps is performed. First, program data is allocated to memory resources. Array variables are mapped to RAMs (internal or external) and some scalar variables to registers in the array. By default, a distinct internal memory is assigned to each array variable. Arrays are mapped to external memories, if the size of an array oversizes the size of each internal memory, or if the programmer explicitly guides the compiler for such mapping (by using special pragma directives).

Next, the CDFG, which directly reflects the NML structure needed, is generated. One element of a PAE is allocated for each operator in the CDFG. Straight-line code without array references is directly mapped to a set of PAE elements connected according to the dataflow. Because of the data-driven mechanism, no explicit control or scheduling is needed. The same is true for conditional execution. Both branches of an IF statement are executed in parallel and MUX operators select the correct output depending on the condition. Fig. 9.3(a) shows a simple conditional statement and its CDFG. IF statements containing array references or loops are implemented with DEMUX operators. They forward data only to the branch selected by the IF condition. The outputs of both branches are connected to the subsequent PAE elements. With this implementation, only selected branches receive data and execute and thus no conflict occurs. Fig. 9.3(b) shows the conditional statement of Fig. 9.3(a) implemented this way.

Accesses to the same memory have to be controlled explicitly to maintain the correct execution order. MERGE operators forward addresses and data in the correct order to the RAM, and DEMUX operators forward values read to the correct subsequent operator. Finite state machines for generating the correct sequence of events (to control these operators) are synthesized by the

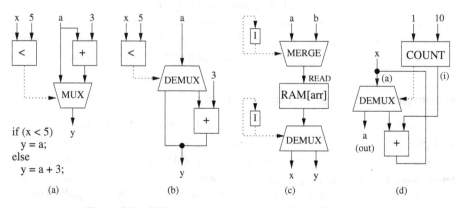

Figure 9.3. XPP array structures mapped from C programs.

compiler. I/O port accesses are handled in a similar way. Fig. 9.3(c) shows the implementation of the statements x = arr[a]; y = arr[b];. In this example, the events controlling the MERGE and DEMUX operators are simply generated by an event register with a feedback cycle. The register is preloaded with "0" and is configured to invert its output value.

Next, we discuss loops. All scalar variables updated in the loop body are treated as follows. In all but the last iteration, a DEMUX operator routes the outputs of the loop body back to the body inputs. Only the results of the last iteration are routed to the loop output. The events controlling the DEMUX are generated by the loop counter of FOR loops or by the comparator evaluating the exit condition of WHILE loops. Note that the internal operators' outputs cannot just be connected to subsequent operators since they produce a result in each loop iteration. The required last value would be hidden by a stream of intermediate values. As opposed to registers in FPGAs, the handshaking protocol requires that values in XPP registers are consumed or explicitly discarded before new values are accepted. Fig. 9.3(d) shows the (slightly simplified) CDFG for the following C code: a = x; for (i = 1; i < = 10; i++) a = a + i; Note that the counter generates ten data values (1.10), but eleven events. After ten loop iterations a "1" event routes the final value of 'a' to the output.

If the loop body contains array or I/O accesses, a loop iteration may only start after the previous iteration has terminated since the original access order must be maintained. The compiler generates events enforcing this. For generating more efficient XPP configurations, MODGen generates pipelined operator networks for inner loops which have been annotated for vectorization by the preprocessing phase. In other words, subsequent loop iterations are started before previous iterations have finished. Data flows continuously through the operator pipelines. By applying pipelining balancing techniques, maximum throughput is achieved. Note that registers in interconnection segments or in the FREG and BREG elements are used for delay registers needed for pipelining balancing and thus there is no need to waste ALU elements for this purpose.

To reduce the number of memory accesses, the compiler automatically removes some redundant array reads. Inside loops, loads of subsequent array elements are transformed to a single load and a shift register scheme.

Generating Multiple Configurations

A program too large to fit in an XPP can be handled by splitting it in several parts (configurations) such that each one is mappable [3]. Although this approach enables the mapping of large computational structures, other goals shall be considered. Our approach also considers a judicious selection of the set of configurations, such that the overall execution time of the application is minimized [4]. Such minimization can be mainly achieved by the following strategies: (1) reduction of each partition's complexity can reduce the interconnection delays (long connections may require more clock cycles); (2) reduction of the number of references to the same resource, e.g. memory, in each partition, by distributing the overall references among partitions: (3) reduction of the overall configuration overhead by overlapping fetching, configuring and computing of distinct partitions.

In XPP-VC, a C program can be manually mapped to multiple configurations by using annotations. Otherwise, a temporal partitioning algorithm automatically splits the program into mappable partitions. Apart from generating the NML representation of each partition, the compiler generates the NML section specifying the control flow of configurations. Pre-fetch statements can be automatically inserted by the compiler. Since internal memories preserve their state between configurations, they are used to communicate values between configurations.

Bearing in mind the following goals, we have developed and evaluated two temporal partitioning schemes: (a) find mappable solutions (i.e., sets of feasible configurations) [3]; (b) find mappable solutions and try to improve the latency by exploiting the pipelining among configure-execute stages [4]. In (a), the set of configurations is constructed by iteratively merging adjacent AST (abstract syntax tree) nodes, forming larger configurations. The merging is guided by estimations of the required resources and only depends on the feasibility to implement the configuration in the given XPP array. In (b), the merge operation is more selective and besides the feasibility, it is only performed if it leads to better overall performance.

Both strategies only perform loop partitioning if an entire loop cannot be mapped to a single configuration. When this occurs, the algorithms are applied to the body of the loop and the "loop dissevering" technique is used [3]. A similar approach is used when a top level AST cannot be mapped to a single configuration.

To furnish feasible partitions, the approach ultimately uses checks with MODGen and xmap before selecting the final set of partitions. The algorithm avoids intermediate checks with xmap, since it might not be tolerable to check each performed merge. Each time the estimated resources obtained from MODGen exceed the size constraint, the algorithm tries another merge. The approach uses three levels of checks and it iterativelly performs merge steps decreasing the size constraint until a feasible set is found [3].

9.4   Experimental Results

For evaluating the compiler and the explained temporal partitioning schemes, a set of benchmarks is used. They have from 18 to 228 lines of C code, between two and 16 loops, and between two and six array variables. The examples are compiled using temporal partitioning (1) whenever it is required to make feasible the mapping and (2) to improve the overall performance. The target platform is a 32-bit XPP array with 16×18 PAEs (32 of them with memories). In the results we show the total number of PAEs of the array used by a certain configuration despite some cases only use one or two of the PAE elements.

The first example is an implementation of the 8×8 discrete cosine transform (DCT), based on two matrix multiplications. Splitting the program improves the overall latency by 14%, requiring 58% fewer PAEs. When considering full unrolling of each of the innermost loops in the matrix multiplications, splitting the program improves the overall latency by 41%, requiring 52% fewer PAEs. For an implementation of the forward 2D Haar wavelet transform, an increase in performance of 7% and a reduction of 55% in size are achieved with scheme (2). For the Conway's Game of Life, a performance increase of 49% and a reduction of 15% in size are achieved (using 6 instead of 4 configurations). The results show that approach (2) leads to sets of configurations that reduce the reconfiguration overhead and thus may lead to performance gains and to lower requirements as far as the number of resources is concerned.

In order to show the compiler efficiency, achieved speedups with various examples are presented in Table 9.1. The third column shows the CPU time (using a PC with a Pentium III at 933 MHz) to compile each example with XPP-VC (from the C program to the binaries). Columns #Part, #PAEs, and #Max represent the number of partitions, number of PAEs used (in the cases with more than one partition, we show the number of PAEs of the largest configuration and the sum of PAEs for all configurations), and the maximum number of concurrently executed PAEs (as an indication of ILP degree), respectively. Column 7 shows the total number of clock cycles (taking into account setup, fetching, configuring, data communication and computation). The last column shows the speedup obtained with examples executed on the XPP at 150 MHz over the same examples compiled with gcc and running on a PIII at 933 MHz. Using the compiler, speedups from 2.8 to 98 have been achieved. Since, as far as the

Table 9.1. Results obtained with XPP-VC.

Example	Loop unr. (#loops)	Compilation time (sec)	#Part	#PAEs	#Max	Total (#ccs)	Speedup
dct	yes (2)	2.2	1	240	26	11,252,333	2.9
fir	yes (1)	0.9	1	132	67	16,853	40.0
bpic	yes (3)	3.4	1	237	12	1,047,733	2.8
edge_det	yes (2)	5.6	5	54/225	12	3,541,818	9.3
fdct	no	1.8	2	150/299	14	1,240,078	12.7
median	yes (3)	1.3	2	201/231	44	28,090	98.0

clock frequency is concerned, the PIII's is 6.2 times faster than the considered XPP, the results are very promising. Note that there are other advantages to use the XPP instead of a microprocessor (e.g., lower energy consumption).

For all examples, the compilation from the source program to the binary configuration file takes only a few seconds. This shows that it is possible, when targeting coarse-grained reconfigurable architectures, to have compilation times comparable to the ones achieved by software compilation, and to still accomplish efficient implementations.

9.5   Related Work

The work on compiling high-level descriptions onto reconfigurable logic has been the focus of many researchers since the first simple attempts (e.g., [9]). Not so many compilers have considered the integration of temporal partitioning. Probably, the first compiler to include temporal partitioning has been the dbC Napa compiler [10]. In that approach, each function defines a configuration and the only way to control the generation of configurations is to manually group program statements in functions. An automatic approach is used in the Galadriel & Nenya compilers [11].

Other authors compile to architectures, which explicitly support hardware virtualization (e.g., PipeRench [12]). However, they only support acyclic computations or computations where cyclic structures can be transformed to acyclic ones.

Most of the current approaches attempt to achieve a minimum number of configurations (e.g., [13]). Those schemes only consider another partition after the current one uses as much available resources as possible, and they are not aware to the optimization that could be applied to reduce the overall execution by overlapping the fetching, configuring and computing steps. One of the first attempts to reduce the configuration overhead in the context of temporal partitioning has been presented in [14].

Although most compilers target fine-grained architectures (e.g., FPGAs), compilers to coarse-grained reconfigurable architectures are becoming focus of research work (e.g., the compiler of SA-C programs to Morphosys [15]). Unlike other compilers, the research efforts on the XPP-VC compiler include schemes to deal with large programs and to efficiently deal with the configure-execute paradigm. *Loop dissevering*, a method for temporally partitioning any kind of loop, has been also proposed in the context of the XPP-VC [3].

9.6   Conclusions

This chapter has described a compiler to map programs in a C subset to the XPP architecture. Combined with a fast place and route tool, it provides

114

a complete "push-button" path from algorithms to XPP binaries within short compilation times. Since the XPP includes a configuration manager, programs are compiled to a single, self-contained binary file, that can be directly executed on the XPP. Such file includes data for configurations and the programming code to control the configuration manager.

Supported by temporal partitioning, the approach enables the mapping of programs requiring more than the available resources on a particular XPP, and attempts to furnish implementations with maximum performance by hiding some of the reconfiguration time. The obtained results prove that a judicious selection of configurations can reduce the overall execution time, by furnishing solutions that overlap execution stages. Considering the configure-execute model, the results show that in many cases a smaller array can be used without sacrificing performance.

Ongoing work focuses on tuning the estimation steps, on optimizations, on removing the current restrictions for pipelining loops, and on supporting other loop transformations.

References

[1] R. Hartenstein, "A Decade of Reconfigurable Computing: a Visionary Retrospective," in Proc. Design, Automation and Test in Europe (DATE'01), 2001, pp. 642–649.

[2] V. Baumgarte, et al., "PACT XPP—A Self-Reconfigurable Data Processing Architecture," in The Journal of Supercomputing, Kluwer Academic Publishers, Sept. 2003, pp. 167–184.

[3] J. M. P. Cardoso, and M. Weinhardt, "XPP-VC: A C Compiler with Temporal Partitioning for the PACT-XPP Architecture," in Proc. 12th Int'l Conference on Field Programmable Logic and Applications (FPL'02), LNCS 2438, Springer-Verlag, 2002, pp. 864–874.

[4] J. M. P. Cardoso, and M. Weinhardt, "From C Programs to the Configure-Execute Model," in Proc. Design, Automation and Test in Europe (DATE'03), Munich, Germany, March 3–7, 2003, pp. 576–581.

[5] M. Weinhardt, and W. Luk, "Pipeline Vectorization," in IEEE Transactions on Computer-Aided Design of Integrated Circuits and Systems, Feb. 2001, pp. 234–248.

[6] R. Hartenstein, R. Kress, and H. Reining, "A Dynamically Reconfigurable Wavefront Array Architecture for Evaluation of Expressions," in Proc. Int'l Conference on Application-Specific Array Processors (ASAP'94), 1994.

[7] SUIF Compiler system, "The Stanford SUIF Compiler Group," http://suif.stanford.edu

[8] S. S. Muchnick, Advanced Compiler Design and Implementation. Morgan Kaufmann Publishers, Inc., San Francisco, CA, USA, 1997.

[9] I. Page, and W. Luk, "Compiling occam into FPGAs," in FPGAs, Will Moore and Wayne Luk, eds., Abingdon EE & CS Books, Abingdon, England, UK, 1991, pp. 271–283.

[10] M. Gokhale, and A. Marks, "Automatic Synthesis of Parallel Programs Targeted to Dynamically Reconfigurable Logic Array," in Proc. 5th Int'l Workshop on Field Programmable Logic and Applications (FPL'95), LNCS, Springer-Verlag, 1995, pp. 399–408.

[11] J. M. P. Cardoso, and H. C. Neto, "Compilation for FPGA-Based Reconfigurable Hardware," in IEEE Design & Test of Computers Magazine, March/April, 2003, vol. 20, no. 2, pp. 65–75.

[12] S. Goldstein, et al., "PipeRench: A Reconfigurable Architecture and Compiler," in IEEE Computer, Vol. 33, No. 4, April 2000, pp. 70–77.

[13] I. Ouaiss, et al., "An Integrated Partioning and Synthesis System for Dynamically Reconfigurable Multi-FPGA Architectures," in Proc. 5th Reconfigurable Architectures Workshop (RAW'98), Orlando, Florida, USA, March 30, 1998, pp. 31–36.

[14] S. Ganesan, and R. Vemuri, "An Integrated Temporal Partitioning and Partial Reconfiguration Technique for Design Latency Improvement," in Proc. Design, Automation & Test in Europe (DATE'00), Paris, France, March 27–30, 2000, pp. 320–325.

[15] G. Venkataramani, et al., "Automatic compilation to a coarse-grained reconfigurable system-on-chip," in ACM Transactions on Embedded Computing Systems (TECS), Vol. 2, Issue 4, November 2003, pp. 560–589.

Chapter 10

Run-time Defragmentation for Dynamically Reconfigurable Hardware[*]

Manuel G. Gericota,[1] Gustavo R. Alves,[1] Miguel L. Silva,[2] and José M. Ferreira[2,3]

[1] *Department of Electrical Engineering—ISEP*
Rua Dr. António Bernardino de Almeida, 431
4200-072 Porto, Portugal
{mgg, galves}@dee.isep.ipp.pt

[2] *Department of Electrical and Computer Engineering—FEUP*
Rua Dr. Roberto Frias
4200-465 Porto, Portugal
mlms@fe.up.pt

[3] *Institutt for datateknikk—Høgskolen i Buskerud*
Frogsvei 41
3601 Kongsberg, Norway
jmf@fe.up.pt

Abstract   Reconfigurable computing experienced a considerable expansion in the last few years, due in part to the fast run-time partial reconfiguration features offered by recent SRAM-based Field Programmable Gate Arrays (FPGAs), which allowed the implementation in real-time of dynamic resource allocation strategies, with multiple independent functions from different applications sharing the same logic resources in the space and temporal domains.

However, when the sequence of reconfigurations to be performed is not predictable, the efficient management of the logic space available becomes the greatest challenge posed to these systems. Resource allocation decisions have to be made concurrently with system operation, taking into account function priorities and optimizing the space currently available. As a consequence of the

[*] This work was supported by *Fundação para a Ciência e Tecnologia* under contract number POCTI/ 33842/ESE/2000

P. Lysaght and W. Rosenstiel (eds.),
New Algorithms, Architectures and Applications for Reconfigurable Computing, 117–129.
© 2005 *Springer. Printed in the Netherlands.*

unpredictability of this allocation procedure, the logic space becomes fragmented, with many small areas of free resources failing to satisfy most requests and so remaining unused.

A rearrangement of the currently running functions is therefore necessary, so as to obtain enough contiguous space to implement incoming functions, avoiding the spreading of their components and the resulting degradation of system performance. A novel active relocation procedure for Configurable Logic Blocks (CLBs) is herein presented, able to carry out online rearrangements, defragmenting the available FPGA resources without disturbing functions currently running.

Keywords: Reconfigurable computing, partial and dynamic reconfiguration, active function, dynamic relocation, fragmentation, defragmentation

Introduction

The idea of using a reconfigurable device to implement a number of applications requiring more resources than those currently available emerged in the beginning of the nineteen-nineties. The concept of virtual hardware, defined in [Long, 1993], pointed to the use of temporal partitioning as a way to implement those applications whose area requirements exceed the reconfigurable logic space available, assuming, to a certain extent, the availability of unlimited hardware resources. The static implementation of a circuit was split up into two or more independent hardware contexts, which could be swapped during runtime [Hauck, 1997], [Trimberger, 1998], [Kaul, 1999], [Cardoso, 1999], [Cantó, 2001]. Extensive work was done to improve the multi-context handling capability of these devices, by storing several configurations and enabling quick context switching [Singh, 2000], [Maestre, 2001], [Meiβner, 2001], [Sanchez-Elez, 2002]. The main goal was to improve the execution time by minimising external memory transfers, assuming that some amount of on-chip data storage was available in the reconfigurable architecture. However, this solution was only feasible if the functions implemented on hardware were mutually exclusive on the temporal domain, e.g. context-switching between coding/decoding schemes in communication, video or audio systems; otherwise, the length of the reconfiguration intervals would lead to unacceptable delays in most applications.

These restrictions have been overtaken by higher levels of integration, due to the employment of very deep submicron and nanometer scales, and by the use of higher frequencies of operation. The increasing amount of logic available in FPGAs and the reduction of the reconfiguration time, partly owing to the possibility of partial reconfiguration, extended the concept of virtual hardware to the implementation of multiple applications sharing the same logic resources in the temporal and spatial domains.

The main goal behind the temporal and spatial partitioning of reconfigurable logic resources is to achieve the maximum efficiency of reconfigurable systems, pushing up resource usage and taking the maximum advantage of its flexibility [Diessel, 2000]. This approach is viable because an application is composed of a set of functions, predominantly executed sequentially, or with a low degree of parallelism, whose simultaneous availability is not required. On the other hand, partial reconfiguration times are sufficiently small, in the order of a few milliseconds or less, depending on the configuration interface and on the complexity of the function being implemented, for those functions to be swapped in real time. If a proper floorplanning schedule is devised, dynamic function swapping enables a single device to run a set of applications, which in total may require more than 100% of the FPGA resources.

Reconfiguration time has decreased consistently from one generation to the next, but it may still be too long, when compared with the function execution period. However, its reduction to virtually zero may be achieved by employing prefetching techniques, similar to those used in general purpose computer systems [Li, 2002]. Functions whose availability is about to be required are configured in advance, filling up unused resources or resources previously occupied by functions that are no longer needed. This situation is illustrated in Figure 10.1, where a number of applications share the same reconfigurable logic space in both the temporal and spatial domains. The earlier execution of the reconfiguration task of a new incoming function during the float interval (see Figure 10.1) assures its readiness when requested by the corresponding application flow [Jeong, 2000]. However, this will be true only if the float interval is greater than or equal to the reconfiguration interval required by the new incoming function. An increase in the degree of parallelism may retard the reconfiguration of incoming functions, due to lack of space in the FPGA. Consequently, delays will be introduced in the application's execution time, systematically or not, depending on the application flow.

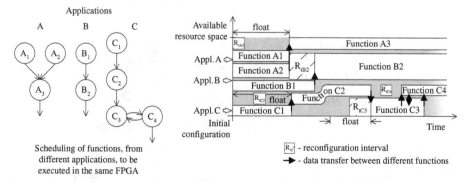

Figure 10.1. Temporal scheduling of functions from different applications in the temporal and spatial domains.

120

The aim of the proposed approach is to allocate to each function as many resources as it needs to execute independently of all the others, as if it were part of a single application running on a chip just large enough to support it. However, several problems have to be solved to achieve this objective.

An incoming function may require the relocation of other functions already implemented and running, in order to release enough contiguous space for its configuration (see function C2 in Figure 10.1). Notice also that spreading the components of an incoming function by the available resources would lead to a degradation of its performance, delaying tasks from completion and reducing the utilisation of the FPGA.

Every multiple independent functions that share the logic space have unique spatial and temporal requirements. Over time, as they are loaded and unloaded, the spatial distribution of the unoccupied area is likely to become fragmented, similar to what occurs in memory management systems when RAM is allocated and deallocated. These portions of unallocated resources tend to become so small that they fail to satisfy any request and so remain unused [Compton, 2002]. This problem is illustrated in Figure 10.2 in the form of a 3-D floorplan and function schedule [Vasilko, 1999], where each shadow area corresponds to the optimal space occupied by the implementation of a single function.

If the requirements of functions and their sequence are known in advance, suitable arrangements can be designed, and sufficient resources can be provided to handle them in parallel [Teich, 1999]. However, when placement decisions have to be made online, it is possible that a lack of contiguous free

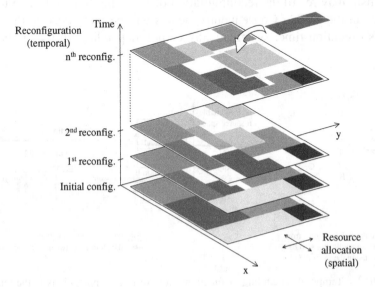

Figure 10.2. 3-D representation of the fragmentation problem.

resources prevents functions from being implemented, even when the total number of resources available is sufficient. In these cases, an increase on the available resources is a poor solution, since it decreases the efficiency of the system.

The problem described may be solved by an efficient online management of the available resources. If a new function cannot be implemented immediately due to a lack of contiguous free resources, a suitable rearrangement of a subset of the functions currently running might solve the problem. Three methods are proposed in [Diessel, 2000] to find such (partial) rearrangements, so as to increase the rate at which resources are allocated to waiting functions, while minimising disruptions on running functions that are to be relocated. However, the disruption of those functions suspends the execution of their owner applications and, eventually, of the whole system operation. Therefore, this proposal is only feasible if the implementation of the rearrangement does not stop the execution of the functions that are to be moved.

To deal with this problem, a new concept is introduced—the dynamic relocation—defined as the possibility of performing the partial or total relocation of a function even if it is active, i.e. if the function to be moved is currently being used by an application [Gericota, 2002]. This concept enables the defragmentation of the FPGA logic space online, gathering small fragments of unused resources in areas large enough for new incoming functions to be set up, without disturbing active functions. Based on new partial and dynamically reconfigurable FPGA features, a procedure to implement this concept is presented in the next sections.

Dynamic Relocation

Conceptually, an FPGA may be described as an array of uncommitted CLBs, surrounded by a periphery of IOBs (Input/Output Blocks), which can be interconnected through configurable routing resources. The configuration of all these elements is controlled by an underlying set of memory cells, as illustrated in Figure 10.3.

Any online management strategy implies the existence of a transparent relocation mechanism, whereby a CLB currently being used by a given function has its current functionality transferred into another CLB, without disturbing the system operation. This relocation mechanism does more than just copying the functional specification of the CLB to be relocated: the corresponding interconnections with the rest of the circuit have also to be reestablished; additionally, according to its current functionality, it may also be necessary to transfer its internal state information. This transference is one of the two major issues posed to the transparent relocation of the functionality of a CLB. The other issue is related to the organization of the configuration memory.

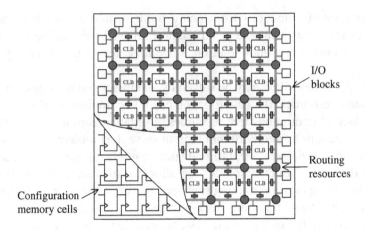

Figure 10.3. Conceptual representation of an FPGA.

The configuration memory may be visualised as a rectangular array of bits, which are grouped into one-bit wide vertical frames extending from the top to the bottom of the array. A frame is the smallest unit of configuration that can be written to or read from the configuration memory. Frames are grouped together into larger units called columns. To each CLB column corresponds a configuration column with multiple frames, where internal CLB configuration and state information are mixed with column routing and interconnect information. The partition of the entire FPGA configuration memory into frames enables its partial reconfiguration, which is carried out sequentially over one or more CLB configuration columns. The rewriting of the same configuration data is completely harmless, and does not generate any transient signals.

When the functionality of a given CLB is dynamically relocated, even into a CLB belonging to the same column, more than one column may be involved, since the CLB input and output signals (as well as those on its replica) may cross several columns before reaching its source or destination. Due to the sequential nature of the configuration mechanism, signals from the original CLB may be broken before being totally reestablished from its replica, disturbing or even halting its operation. Moreover, the functionality of the replica CLB must be perfectly stable before its outputs are connected to the system, preventing output hazards. A set of experiments performed with a Virtex XCV200 [Xilinx, 2002] FPGA demonstrated that the only possible solution was to divide the relocation procedure in two phases [Gericota, 2002], as illustrated in Figure 10.4 (for reasons of intelligibility, only the relevant interconnections are represented).

In the first phase, the internal configuration of the CLB is copied into its new location and the inputs of both CLBs are placed in parallel. Due to the slowness of the reconfiguration procedure, when compared with the speed of operation of current applications, the outputs of the replica CLB are already perfectly stable

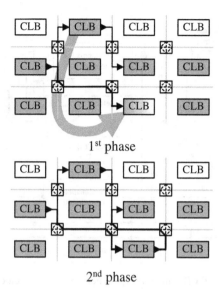

1st phase

2nd phase

Figure 10.4. Two-phase CLB relocation procedure.

when they are connected to the circuit in the second phase. To avoid output instability, both CLBs (the original and its replica) must remain in parallel for at least one clock cycle. Afterwards, and for the same reason, the original CLB inputs may only be disconnected after the outputs are disconnected; otherwise, changes produced in the input signals may extemporaneously be reflected in the outputs. As the rewriting of the same configuration data does not generate any transient signals, this procedure does not affect the remaining resources covered by the rewriting of the configuration frames required to carry out the relocation procedure of a CLB.

If the current CLB function is purely combinational, this two-phase relocation procedure will be sufficient. However, in the case of a sequential function, successful completion of the procedure described also implies the correct transference of the internal state information. This information must be preserved and no update operations may be lost while the CLB functionality is being relocated. The solution to this problem depends on the type of implementation of the sequential circuit: synchronous free-running clock circuits, synchronous gated-clock circuits, and asynchronous circuits.

When dealing with synchronous free-running clock circuits, the two-phase relocation procedure is also effective. Between the first and the second phases, the replica CLB receives the same input signals as the original CLB, meaning that all their flip-flops (FFs) acquire the same state information at the active transition of the clock signal. The experimental replication of CLBs with FFs driven by a free-running clock confirmed the effectiveness of this method.

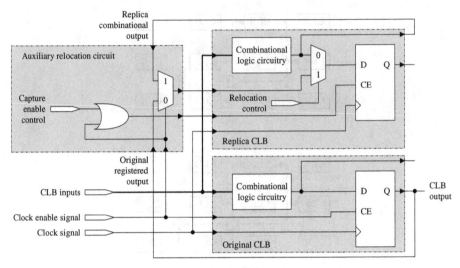

Figure 10.5. CLB relocation for synchronous gated-clock circuits.

No loss of state information or the presence of output hazards was observed [Gericota, 2002].

When using synchronous gated-clock circuits, where input acquisition by the FFs is controlled by the state of the clock enable signal, the previous method does not ensure that the correct state information is captured by the replica CLB, since that signal may not be active during the relocation procedure. Besides, to set it as part of this procedure is incorrect, because the value present at the input of the replica CLB FFs may change in the meantime, and a coherency problem would occur. Therefore, it is necessary to use an auxiliary relocation circuit capable of ensuring the correct transference of the state information from the original CLB FFs to the replica CLB FFs, while enabling its update by the circuit, at any moment, without losing new state information or delaying the relocation procedure. The whole relocation scheme is represented in Figure 10.5. Only one logic cell is shown (each CLB comprises several of such blocks), not only for reasons of simplicity, but also because each CLB cell can be considered individually.

The temporary transfer paths established between the original CLB logic cells and their replicas do not affect their functionality, since only free routing resources are used. No changes in the CLB logic cell structure are required to implement this procedure. The OR gate and the 2:1 multiplexer that are part of the auxiliary relocation circuit must be implemented during the relocation procedure in a nearby (free) CLB.

The inputs of the 2:1 multiplexer present in the auxiliary relocation circuit receive one temporary transfer path from the output of the original CLB FF and another from the output of the combinational logic circuitry in the replica CLB (which is normally applied to the input of its own FF). The clock enable signal of the original CLB FF also controls this multiplexer while the relocation procedure is carried out. If this signal is not active, the output of the original CLB FF is applied to the input of the replica CLB FF. The capture enable control signal, generated by the auxiliary relocation circuit, is activated and accordingly the replica FF stores the transferred value at the active transition of the clock signal, acquiring the state information present in the original CLB FF. If the clock enable signal is active, or is activated during this process, the multiplexer selects the output of the combinational circuitry of the replica CLB and applies it to its own FF input, which is updated at the same clock transition and with the same value as the original CLB FF, therefore guaranteeing state coherency. The flow diagram of the relocation procedure is presented in Figure 10.6.

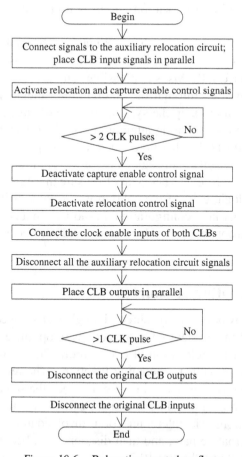

Figure 10.6.   Relocation procedure flow.

The capture enable control signal and the relocation control signal are driven by configuration memory bits. The latter directs the state information value to the input of the replica CLB FF, while the capture enable control signal enables its acquisition. It is therefore possible to control the whole relocation procedure through the same interface used for normal reconfiguration, and thus no extra resources are required for this purpose.

The relocation of a CLB is performed individually, between the original CLB and its replica (more precisely, between each corresponding CLB logic cell), even if many blocks were relocated simultaneously. Therefore, this approach is also applicable to multiple clock/multiple phase applications, without requiring any modification.

This method is also effective when dealing with asynchronous circuits, where transparent data latches are used instead of FFs. In this case, the control enable signal is replaced in the latch by the input control signal, with the value present in the input of the latch being stored when the control signal changes from '1' to '0'. The same auxiliary relocation circuit is used and the same relocation sequence is followed.

The Look-Up Tables (LUTs) in the CLB may also be configured as memory modules (RAMs) for user applications. However, the extension of this concept to the relocation of LUT/RAMs is not feasible. The content of the LUT/RAMs may be read and written through the configuration memory, but there is no mechanism, other than to stop the system, capable of ensuring the coherency of the values, if there is a writing attempt during the relocation procedure, as stated in [Huang, 2001]. Furthermore, since frames span an entire column of CLBs, the same given bit in all logic cells of the column is written with the same command. Therefore, it is necessary to ensure that either all the remaining data in the whole logic cells of the column is constant, or it is also modified externally through partial reconfiguration. In short, even if not being replicated, LUT/RAMs should not lie in any column that could be affected by the relocation procedure.

Rearranging Routing Resources

Due to the scarcity of routing resources, it might also be necessary to perform a rearrangement of the existent interconnections to optimise the occupancy of such resources after the relocation of one or more CLBs, and to increase the availability of routing paths to incoming functions. The relocation of routing resources does not pose any special problems, since the same concept used in the two-phase relocation procedure of active CLBs is valid to relocate local and global interconnections. The interconnections involved must first be duplicated to establish an alternative path, and then disconnected, becoming available to be reused, as illustrated in Figure 10.7.

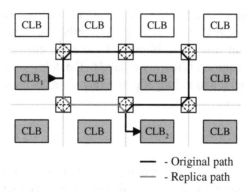

Figure 10.7. Relocation of routing resources.

However, since different paths are used while paralleling the original and replica interconnections, each path will most likely have a different propagation delay. Consequently, if the signal level at the output of the CLB source changes, the signal at the input of the CLB destination will exhibit an interval of fuzziness, as shown in Figure 10.8. Notwithstanding, the impedance of the routing switches will limit the current flow in the interconnection, and hence this behaviour does not damage the FPGA. For that reason, the propagation delay associated to the parallel interconnections shall be the longer of the two paths [Gericota, 2003].

The dynamic relocation of CLBs and interconnections should have a minimum influence (preferably none) in the system operation, as well as reduced overhead in terms of reconfiguration cost. The placement algorithms (in an attempt to reduce path delays) gather in the same area the logic that is needed to implement the components of a given function. It is unwise to disperse it, since it would generate longer paths (and hence, an increase in path delays). On the other hand, it would also put too much stress upon the limited routing resources. Therefore, the relocation of the CLBs should be performed between nearby CLBs. If necessary, the relocation of a complete function may take place in several stages to avoid an excessive increase in path delays during the relocation interval.

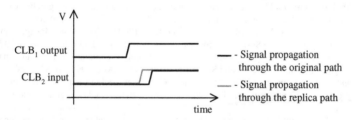

Figure 10.8. Propagation delay during the relocation of routing resources.

The reconfiguration cost depends on the number of reconfiguration frames needed to relocate each CLB, since a great number of frames would imply a longer rearrangement time. The impact of the relocation procedure in those functions currently running is mainly related to the delays imposed by rerouted paths, because it might lead to longer paths, therefore decreasing the maximum frequency of operation.

Conclusion

An efficient strategy for the reconfigurable logic space management is fundamental to achieve maximum performance in a reconfigurable system. The online rearrangement of currently implemented functions is therefore of vital importance to enable a proper defragmentation of the logic space and the prompt implementation of incoming functions.

The proposed procedure allows the dynamic relocation of active functions and enables the implementation of rearrangement algorithms without producing any time overhead or disturbing the applications currently running. As a result, the online dynamic scheduling of tasks in the spatial and temporal domains becomes feasible in practice, enabling several applications—that altogether may require more than 100% of the FPGA resources—to share the same hardware platform in a completely transparent way.

References

Cantó, E., J. M. Moreno, J. Cabestany, I. Lacadena, and J. M. Insenser. (2001). "A Temporal Bipartitioning Algorithm for Dynamically Reconfigurable FPGAs," IEEE Trans. on VLSI Systems, Vol. 9, No. 1, Feb. 2001, pp. 210–218.

Cardoso, J. M. P., and H. C. Neto. (1999). "An Enhanced Static-List Scheduling Algorithm for Temporal Partitioning onto RPUs," Proc. 10th Intl. Conf. on VLSI, pp. 485–496.

Compton, K., Z. Li, J. Cooley, S. Knol, and S. Hauck. (2002). "Configuration, Relocation and Defragmentation for Run-Time Reconfigurable Computing," IEEE Trans. on VLSI Systems, Vol. 10, No. 3, June 2002, pp. 209–220.

Diessel, O., H. El Gindy, M. Middendorf, H. Schmeck, and B. Schmidt. (2000). "Dynamic scheduling of tasks on partially reconfigurable FPGAs," IEE Proc.-Computer Digital Technology, Vol. 147, No. 3, May 2000, pp. 181–188.

Gericota, M. G., G. R. Alves, M. L. Silva, and J. M. Ferreira. (2002). "On-line Defragmentation for Run-Time Partially Reconfigurable FPGAs." In Glesner, M., Zipf, P., and Renovell, M., editors, Proc. 12th Intl. Conf. Field Programmable Logic and Applications: Reconfigurable Computing is Going Mainstream, Lecture Notes in Computer Science 2438, pp. 302–311. Springer.

Gericota, M. G., G. R. Alves, M. L. Silva, and J. M. Ferreira. (2003). "Run-Time Management of Logic Resources on Reconfigurable Systems," Proc. Design, Automation and Test in Europe, pp. 974–979.

Hauck, S., and G. Borriello. (1997). "An Evaluation of Bipartitioning Techniques," IEEE Trans. on Computer Aided Design of Integrated Circuits and Systems, Vol. 16, No. 8, Aug. 1997, pp. 849–866.

Huang, W., and E. J. McCluskey. (2001). "A Memory Coherence Technique for Online Transient Error Recovery of FPGA Configurations," Proc. 9th ACM Intl. Symp. Field Programmable Gate Arrays, pp. 183–192.

Jeong, B., S. Yoo, S. Lee, and K. Choi. (2000). "Hardware-Software Cosynthesis for Run-time Incrementally Reconfigurable FPGAs," Proc. 2000 Asia South Pacific Design Automation Conf., pp. 169–174.

Kaul, M., and R. Vemuri. (1999). "Temporal Partitioning Combined with Design Space Exploration for Latency Minimization of Run-Time Reconfigured Designs," Proc. Design, Automation and Test in Europe, pp. 202–209.

Li, Z., and S. Hauck. (2002). "Configuration Prefetching Techniques for Partial Reconfigurable Coprocessor with Relocation and Defragmentation," Proc. 10th ACM Int. Symp. Field-Programmable Gate Arrays, pp. 187–195.

Long, X. P., and H. Amano. (1993). "WASMII: a Data Driven Computer on a Virtual Hardware," Proc. 1st IEEE Workshop on FPGAs for Custom Computing Machines, pp. 33–42.

Maestre, R., F. J. Kurdahi, R. Hermida, N. Bagherzadeh, and H. Singh. (2001). "A Formal Approach to Context Scheduling for Multicontext Reconfigurable Architectures," IEEE Trans. on VLSI Systems, Vol. 9, No. 1, Feb. 2001, pp. 173–185.

Meißner, M., S. Grimm, W. Straßer, J. Packer, and D. Latimer. (2001). "Parallel Volume Rendering on a Single-Chip SIMD Architecture," Proc. IEEE Symp. on Parallel and Large-data Visualization and Graphics, pp. 107–113.

Sanchez-Elez, M., M. Fernandez, R. Maestre, R. Hermida, N. Bagherzadeh, and F. J. Kurdahi. (2002). "A Complete Data Scheduler for Multi-Context Reconfigurable Architectures," Proc. Design, Automation and Test in Europe, pp. 547–552.

Singh, H., M.-H. Lee, G. Lu, F. J. Kurdahi, N. Bagherzadeh, and E. M. C. Filho. (2000). "MorphoSys: An Integrated Reconfigurable System for Data-Parallel and Computation-Intensive Applications," IEEE Trans. on Computers, Vol. 49, No. 5, May 2000, pp. 465–481.

Teich, M., S. Fekete, and J. Schepers. (1999). "Compile-time optimization of dynamic hardware reconfigurations," Proc. Intl. Conf. on Parallel and Distributed Processing Techniques and Applications, pp. 1097–1103.

Trimberger, S. (1998). "Scheduling designs into a time-multiplexed FPGA," Proc. Int. Symp. Field Programmable Gate Arrays, pp. 153–160.

Vasilko, M. (1999). "DYNASTY: A Temporal Floorplanning Based CAD Framework for Dynamically Reconfigurable Logic Systems," Proc. 9th Intl. Workshop on Field-Programmable Logic and Applications, pp. 124–133.

"The Programmable Logic Data Book," Xilinx, Inc., 2002.

Chapter 11

Virtual Hardware Byte Code as a Design Platform for Reconfigurable Embedded Systems

S. Lange, U. Kebschull

Department of Computer Science
University of Leipzig
Augustusplatz 10-11, 04109 Leipzig, Germany
langes@informatik.uni-leipzig.de
kebschull@ti-leipzig.de

Abstract   Reconfigurable hardware will be used in many future embedded applications. Since most of these embedded systems will be temporarily or permanently connected to a network, the possibility to reload parts of the application at run time arises. In the 90ies it was recognized, that a huge variety of processors would lead to a tremendous amount of binaries for the same piece of software. For the hardware parts of an embedded system, the situation today is even worse. The java approach based on a java virtual machine (JVM) was invented to solve the problem for software. In this paper, we show how the hardware parts of an embedded system can be implemented in a hardware byte code, which can be interpreted using a virtual hardware machine running on an arbitrary FPGA. Our results show that this approach is feasible and that it leads to fast, portable and reconfigurable designs, which run on any programmable target architecture.

11.1   Introduction

With a widespread use of reconfigurable hardware such as FPGAs and PLDs in devices as well as the interconnection between them through networks new possibilities and demands arise. It is now possible to not only change software components at runtime, but also the hardware logic itself. This can be done at the customer site without the need of physical access to the device, possibly even without an explicit interaction with the user. In order to allow for a multitude of devices to interact with each other, hardware components should be interchangeable in a way that is abstracting from

131

P. Lysaght and W. Rosenstiel (eds.),
New Algorithms, Architectures and Applications for Reconfigurable Computing, 131–143.
© 2005 *Springer. Printed in the Netherlands.*

the specific design of the underlying hardware architecture. However, due to the lack of a standardized interface to the FPGAs and other common programmable devices, it is problematic to define logic in a general and portable fashion.

Reconfiguration in hardware is synonymous with the incorporation of FPGAs in the design. Unfortunately, the configuration file formats of different FPGAs, which represent the very essence of reconfigurable hardware, differ greatly and prove incompatible. The incompatibilities are not only vendor induced, but moreover root in the different physical layouts of FPGAs. It is therefore not foreseeable that a standardized, compatible format will be available in the future, leaving reconfigurable hardware in the same dilemma, software was in a decade ago. At that time Java was invented to eliminate the compatibility issues caused by the variety of incompatible processor platforms. Ever since its introduction the Java concept has been inarguably successful. Reconfigurable hardware faces the same principle problem as software did. In consequence the Java approach of abstracting software from the underlying processor platform should be applied to reconfigurable hardware. One such solution will be presented in the remainder of this paper.

The approach of the Virtual Hardware Byte Code (VHBC) as described in this paper provides a means to combine the virtues of hardware design with the flexibility inherent in software. Algorithms are designed as hardware but mapped to a special hardware architecture called the Virtual Hardware Machine (VHM) by a dedicated hardware byte code compiler. The VHM is designed to be easily implemented on virtually any underlying hardware substrate and acts as a mediator between an abstract algorithmic description and the hardware itself.

Another important aspect of a design platform for embedded systems is the driving need for low power devices. With the advent of wireless communication devices and hand-held computers (PDA's, cellular phones) a new door has been opened towards computing power in the light of information whenever and wherever it is needed. Fast execution is paramount to keep up with new ever more involved algorithms. However, this when solved entirely in software on state of the art all-purpose processing elements requires ever higher clock frequencies which puts a heavy burden on the energy resources of any hand-held device or embedded system. Using dedicated hardware resources effectively increases the computational power as well as lowers the power consumption, but this comes at the price of very limited flexibility. The VHBC is specifically designed to represent hardware. Designs mapped to the VHM will thus run more effectively than corresponding solutions in pure software on general purpose processors, yielding higher performance and lower power consumption.

11.1.1 State of the Art

At present, the design of a new embedded system usually leads to the deployment of two coexisting design platforms. On one hand, functionality is cast into hardware by means of ASIC design. ASICs provide the highest performance and the lowest power consumption. However, they do not account for later changes in functionality demands making reconfigurability impossible. Moreover, the cost of designing ASICs is skyrocketing. Furthermore, manufacturing techniques have ventured deep into the fields of sub-micron physics, introducing effects unaccounted for in years passed, such as signal integrity, power leakage or electromigration, which in itself makes it harder to design working chips and thus extends the amount of money as well as time spent on designing ASICs. A detailed discussion of these effects, however, is far beyond the scope of this paper and is discussed in detail in the literature [1]. The other design platform mentioned is software. Software, being an abstract description of functionality is inherently exchangeable. It allows not only to account for changing demands, which greatly shortens design cycles, but provides also means to transfer functionality among systems. Furthermore, processors are applied "off the shelf", thus being cost-effective and thoroughly tested. Yet a lot of different processor platforms exist, leading towards compatibility problems and a multitude of equivalent implementations of the same application. Far worse, processors prove to be very power hungry and very complex devices wasting a lot of available computing power because they are designed to be general purpose, but usually only perform very specific tasks when employed in embedded systems.

In conjunction, neither software nor ASIC design can sufficiently solve the problems of modern design challenges. What is needed, is a way to combine the virtues of both design platforms, namely reconfigurability and short design cycles coupled with fast execution and low power consumption, thus making hardware "virtual" [2]. FPGAs were a very important step towards Virtual Hardware, because they offer close resemblance of the performance of custom-built circuits, yet still provide for changing functionality through reconfiguration. However, FPGAs show several disadvantages. Most predominantly, FPGAs do not share a common, standardized way of describing hardware, but rather differ greatly in the layout and format of their bit file descriptions. Furthermore, FPGAs impose harsh limitations towards the size of the Virtual Hardware designs. Although the number of logical gates that can be fit on a FPGA is increasing, it is still too small for a lot of real world designs to be implemented entirely on an FPGA. As another aspect, the time needed to reconfigure a whole FPGA, lies well in the range of several milliseconds to seconds, proving too long for applications which change dynamically during execution. The introduction of partial reconfiguration has helped alleviate the problem, but in consequence leads to a coarse grain place and route—process to bind the partial designs to available resources within the FPGA, which has to be done at runtime on the chip. This however, adds complexity to the surrounding systems, because they have to accommodate for the place and route logic.

Several proposals have been made addressing different shortcomings of current FPGAs. The "Hardware Virtual Machine" [3] [4] project lead by Hugo de Man at the K.U. Leuven gives attention to the problem of incompatible bit files, proposing the definition of an abstract FPGA which provides the essence of FPGAs. Designs are to be mapped onto an abstract FPGA and placed and routed into small fractions. The so mapped and routed fragments pose an abstract yet portable representation of the design. In order to allow specific host FPGAs to make use of it, a Hardware Virtual Machine (HVM) converts the abstract representation to bit files specific to the FPGA and reconfigures it. This approach has the advantage of running the actual design natively on the FPGAs, thus providing high performance. The conversion itself, however, involves a placing of the hardware fragments as well as a routing of signals between them on the FPGA, which requires considerable computational effort and has to be done at design loading time. Furthermore, FPGA architectures differ greatly and their common ground might be too small to account for an efficient representation of Virtual Hardware and allow for great efficiency in the resulting specific bit files.

Another project, "PipeRench" [5] at CMU in Pittsburgh, addresses the spacial limitation imposed by FPGA design. The basic idea is to identify pipeline stages within a given design. Each stage is then treated as a single block of functionality, that is swapped in and out of the FPGA as needed. The control over the reconfiguration process is given to a special purpose processor that transfers the stages from an external memory to the FPGA and vice versa. The project assumes that the reconfiguration of a single pipeline stage can be done with great speed, allowing for stages to execute while others are reconfiguring. The results of this project present a possibility to design Virtual Hardware without the need to consider the spatial extend of the underlying reconfigurable hardware substrate, while also enabling designers to incorporate dynamic reconfiguration code changes into the design, thus clearing the way towards devices that change the functionality provided within the device ad hoc at user request. The pitfalls of the PipeRench approach clearly lie within its requirement of special FPGAs with extremely small reconfiguration delays, as well as the use of a control processor which turns out to be very complex [6]. Furthermore the architecture provides only limited support for interstage feed-back data flow, thus restricting the domain of applications to merely time-pipelinable tasks.

11.1.2   Our Approach

Our approach is different from the aforementioned in that it defines means of implementing the hardware part of an embedded system as an intermediate hardware byte code. This byte code describes the hardware net list on register transfer or logic level. It is compact and provides means to describe the parallelism of the underlying hardware. A Virtual Hardware Machine (VHM) interprets the byte code. The VHM is implemented in VHDL and can therefore be mapped to any target FPGA or even into a CMOS implementation.

The following sections describe the general concept of the Virtual Hardware Byte Code and the Virtual Hardware Machine. To this point a first implementation of the VHM is available, which, however, supports only a subset of all possible features. The byte code reflects this simplification by providing instructions only on the logic level. These restrictions were imposed to allow for a first evaluation of the concept. Nevertheless, it should be evident that the concept is not limited to the current implementation and will in fact be easily adaptable to descriptions on register transfer level, yielding even higher performance.

11.2   The Virtual Hardware Byte Code

The definition of a coding scheme for virtual hardware reveals several important aspects. The most predominant ones are generality and high performance, which unfortunately proof to be quite orthogonal in this matter. Generality for one demands that the vast majority of circuits may be feasible within the execution model of the code while also calling for the highest level of abstraction of the code from the underlying hardware substrate. Performance on the other hand requests the code to be as close to the hardware as possible in order to minimize the overhead induced by emulation on a hardware implementation. Another important aspect is that hardware reveals a very high level of parallelism. This inherent parallelism should map to instruction level parallelism in the code in order to increase performance.

With these goals in mind the Virtual Hardware Byte Code has been defined with a number of features. In order to support a majority of circuits the available instructions mimic basic logic operations and simple RT-level constructs such as adders or multiplexers. This also yields a close resemblance of the code to traditional hardware concepts and thus allows for great efficiency in the actual implementation of an interpreting hardware processor (Virtual Hardware Machine). To account for high performance and alleviate the runtime system from the involved task of extracting instruction level parallelism at runtime the virtual byte code image is pre-scheduled into blocks of instructions which are independent of each other. Hence all instructions belonging to the same block can be executed in parallel without side effects resulting from data dependencies.

Table 11.1. Logic Level Instruction set of the Virtual Hardware Byte Code

OpCode	Mnemonic	Mathematical Expression
0000_2	EOB	N/A
0001_2	NOP	$c = c$
0010_2	MOV	$c = a$
0011_2	NOT	$c = \overline{a}$
0100_2	AND	$c = a \wedge b$
0101_2	NAND	$c = \overline{a \wedge b}$
0110_2	OR	$c = a \vee b$
0111_2	NOR	$c = \overline{a \vee b}$
1000_2	XOR	$c = (\overline{a} \wedge b) \vee (a \wedge \overline{b})$
1001_2	EQ	$c = (a \wedge b) \vee (\overline{a} \wedge \overline{b})$
1010_2	IMP	$c = (a \wedge b) \vee \overline{a}$
1010_2	NIMP	$c = a \wedge \overline{b}$
1100_2	CMOV	$c = \begin{cases} b & \text{if } a \text{ true} \\ c & \text{otherwise} \end{cases}$

The Virtual Hardware Byte Code uses a register transfer model to represent functionality. Instructions have a fixed format and consist of an operation code, two source registers and a destination register. The number of bits addressed by one register has been held flexible to allow for single bit logic instruction as well as multi-bit RTL-operations. The instructions, which are available to manipulate register contents represent basic logic operations as well as data movement instructions. A short overview of the available instruction set for the logic level operations is given in Table 11.1.

The byte code is structured hierarchically. At the topmost level it consists of two major parts, the header section, which provides general information about the circuit embedded in the image as well as information necessary to adapt it to the specific implementation of the Virtual Hardware Machine. The Virtual Hardware itself is encoded in the second part and consists of an arbitrary number of instructions grouped into code blocks. In order to illustrate the rationale behind the available instruction set the CMOV instruction will be described in detail.

The CMOV operation (short for Conditional MOVe) is, as the name suggests, a conditional operation. It moves the content of the second source register to the output register if and only if the value contained in the first source register is not zero. It thus sets itself apart from all other instructions in the instruction set, because the condition introduces a certain amount of control flow to the code and opens the possibility to incorporate a dynamic optimization scheme similar to branch prediction. Conceptually, the CMOV operation takes two data flows, the one which led to the content already stored in the output register and the one which contributes to the second input, and chooses among them according

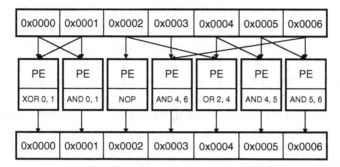

Figure 11.1. Instruction Mapping.

to the condition in the first input register, thereby discarding the result of one data flow. The instructions contained in the data flow which was discarded were thus superfluously computed. Given that the same condition holds in the next cycle it is possible to speculatively prune that data path.

The Virtual Hardware Byte Code defines every position in a code block to be available for every possible type of instruction of the instruction set. Therefore no special positions or slots exist for certain types of instructions. This leaves a degree of freedom in the placement of instructions, which is used to implicitly encode the address of the output register and thereby saves the space otherwise used to explicitly declare it. The address of the destination register is given by the position of the instruction in its code block starting to count at 0. If for example an instruction appears on the third position in the code block its output register address will be 2. Figure 11.1 illustrates how instructions are mapped to output register addresses.

11.3 The Byte Code Compiler

The Byte Code Compiler is a very important feature of the VHBC approach, because it provides the means to compile working hardware designs, coded as a VHDL description, into a portable and efficient VHBC representation, thus removing the need for redesigning working hardware projects. The tool flow within the VHDL compiler can basically be divided into three main stages, the hardware synthesis, the net list to byte code conversion and the byte code optimization and scheduling.

In the first stage the VHDL description is compiled into a net list of standard components and standard logic optimization is performed upon it, resulting in an optimized net list. The design of the compiler chain can be streamlined through the use of off-the-shelf hardware synthesis tools. Current implementations of the VHDL compiler make e.g. use of the FPGAExpress tool from Synopsis. These tools produce the anticipated code using a fairly standardized component library, as in the case of FPGA Express the SimPrim library from Xilinx. The resulting output of the first stage is converted to structural VHDL and passed on to the second stage. Most standard industry VHDL compilers with a support for FPGAs design readily provide the functionality needed for this step and can therefore be applied.

In the second stage the components of the net list are substituted by VHBC fragments to form a VHBC instruction stream. Before, however, the components are mapped to a VHBC representation, the net list is analyzed and optimized for VHBC. The optimization is necessary because commercial compilers targeting FPGAs usually output designs which contain large amounts of buffers to enhance signal integrity otherwise impaired by the routing of the signals. Furthermore, compilers show a tendency towards employing logic representations based on NAND or NOR gates, which are more efficient when cast into silicon. However, the resulting logic structure is more complex, revealing higher levels of logic. The code fragments used for substituting the logic components are based on predefined, general implementations of the latter in VHBC and are adjusted according to the data flow found in the structural description from the first phase, thus registers are allocated and the instructions are sequenced according to the data dependencies inherent.

In the third stage the byte code sequence is optimized and scheduled into blocks of independent instructions. First of all the data flow graph of the entire design is constructed, which is possible due to the lack of control flow instructions such as jumps. The code fragments introduced in the second stage are very general, so the resulting code gives a lot of room to code optimization techniques. One such technique is dead code elimination, which removes unnecessary instructions. The code is further optimized by applying predefined code substitution rules along the data paths, such as XOR extraction or double-negation removal, to reduce the number of instructions and compact the code.

The thus optimized code is scheduled using a list based scheduling scheme [14]. The objective of the scheduling is to group the instructions into code blocks such that the number of code blocks is minimal and the number of instructions per code block is evenly distributed among all code blocks. Furthermore, the time of data not being used, i.e. the number of clock cycles between the calculation of a datum and its use in another operation should be minimal. The scheduled code is then converted to the VHBC image format and the compiler flow concludes.

11.4   The Virtual Hardware Machine

Our approach assumes that hardware descriptions can be translated into a portable byte code which can efficiently be interpreted by a special hardware processor called the Virtual Hardware Machine. The design of the VHM is greatly influenced by the properties of the byte code, namely simple gate level

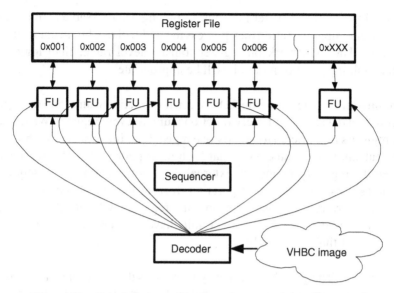

Figure 11.2. Principle components of the Virtual Hardware Machine.

operations and a high level of available instruction level parallelism, which suppose a VLIW-like architecture with a very high number of functional units which possess only very small footprints.

The concept of the VHM is a general one. It aims to be easily adaptable to a variety of of underlying hardware platforms, ranging from standard hardware CMOS implementations to different reconfigurable hardware substrates such as FPGAs. Due to differing platform capabilities VHM implementations differ in the number of available functional units and registers as well as the extend of the available external port capabilities. In principle, the virtual hardware machine consists of five components:

Decoder. The decoder takes the byte aligned instruction input stream and extracts code blocks from it. The instructions embedded in the code blocks are adapted to the specific VHM implementation, thus register addresses might have to be recalculated or the register address sizes possibly need to be enlarged. The adapted instructions are then sent to the instruction caches of the functional units. Furthermore, the decoder is also responsible for resolving problems caused by oversized code blocks, meaning that more instructions are pre-scheduled into a code block than functional units are available. In this case the scheduler tries to split the code blocks into smaller units.

Functional Units. Functional units execute the instructions of the VHBC. In order to allow for an efficient execution, each functional unit contains a processing kernel, sequencer and an instruction cache. The size of the instructions cache differs among implementations.

Register File.   The register file consists of single addressable memory cells. In the current implementation they possess the width of one bit. In later versions, when the VHM will work on register transfer level rather than logic, only registers holding eight or more bit will be appropriate.

Interconnect.   The interconnect between the functional units and the register file allows read access from every functional unit to every register. Write access to the registers is restricted to exactly one register per functional unit, thus every functional unit is associated with a hard-wired output register. The interconnect between the register file and the external ports is triggered by the sequencer. The values present at the input ports are read at the very beginning of each macro cycle, overwriting the corresponding registers, whereas the output port values are altered after all functional units have finished the execution of the instructions of the final code block.

Sequencer.   The global sequencer synchronizes the functional units and triggers the signal propagation to the external ports. Furthermore, it takes care of the reconfiguration of the functional units. Whenever new VHBC images or fragments arrive the sequencer allows the decoder to extract the new instructions and distribute them to the functional units. This can be done, by either stopping the execution of the current VHBC image and fully reconfiguring all FUs with new code, or by inserting or replacing only certain hardware instructions in the instruction caches.

11.5   Results

We have presented the concept of the Virtual Hardware Byte Code in a first preliminary version. To allow for a first feasibility study, the code only facilitates logic operations. Up to now, a rudimentary implementation of the Virtual Hardware Machine in VHDL, a cycle-precise simulation environment as well as a VHDL to VHBC compiler have been implemented to support first evaluations of the concept.

The VHM currently uses up to 16 functional units, with 16 instructions deep I-Caches and 32 registers in the register file. We implemented the VHM using a high level VHDL description and mapped it onto a Xilinx Virtex XCV800 FPGA. First tests show that the current implementation is capable of running with a core speed of at least 100 MHz.

Due to the simplicity of the current implementation three basic designs have been analyzed, a Fulladder (2Add), a 4 bit ripple carry adder (4Add) and a seven segment decoder (7Seg). Furthermore two more involved designs, a 16 bit counter (Count16) and a basic general purpose processor (GPP), were compiled

Table 11.2. Results for different designs running on the VHM and the Xilinx Virtex

			Delay		
	Blocks	Parallelism	VHM	Virtex	Factor
2Add	3	4	30 ns	15.47 ns	1.9
4Add	12	8	120 ns	22.07 ns	5.5
7Seg	8	31	80 ns	12.7 ns	6.2
Count16	15	16	150 ns	18.4 ns	8.2
GPP	37	358	370 ns	58.3 ns	6.3

and simulated using the VHM simulation environment. All five designs show that the claimed high levels of available instruction level parallelism were well grounded. Table 11.2 shows the obtained results. All designs were specified in VHDL and compiled using the VHDL to VHBC compiler. In the table, the column Blocks describes the number of code blocks found in the VHBC code and Parallelism the average number of instruction per block. With the number of code blocks n, a nominal delay d can be calculated for a given design as follows: $d = \frac{n}{100\,MHz} = n \times 10\,ns$. The delay values on the VHM were calculated using this formula. The delays on the Virtex were approximated using the timing information available from the Xilinx VHDL compiler.

In a second experiment the scalability of the approach with respect to different numbers of available FU's was examined. Because the decoder part of the VHM is not in the critical path of the VHM implementation it was excluded from the test design. The design itself was implemented with 2 to 16 functional units and then mapped onto the Xilinx VirtexII. Table 11.3 shows the estimated maximum clock speed, the number of lookup tables occupied by the design as well as the number of slices affected.

The results are quite encouraging to presume further work on the Byte Code, the Virtual Hardware Machine as well as the compiler. They clearly indicate that a design description employing the VHBC performs only factor 5 to 10

Table 11.3. Space and Timing results for different VHMs with different numbers of functional units

#FU	#LUT	#Slices	MAX Clock [MHz]
2	10	5	404
4	36	18	341
8	136	69	283
16	544	272	244

times slower than the same design compiled directly for a specific FPGA, while allowing for portability as well as easy run time reconfiguration without the need for placing and routing. On top of this, the early stage of implementation should be taken into consideration. Code optimizations as well as more sophisticated VHM implementations will definitely show even better performance results.

11.6 Conclusions and Future Work

We have defined a Virtual Hardware Byte Code (VHBC) representation for hardware components in embedded systems, which carries the concept and virtues of Java into the world of hardware design. As a result we received a portable and efficient way to transfer hardware designs via standard network environments. Consequently, we are working on a specially streamlined hardware processor, the Virtual Hardware Machine (VHM), as well as a host of software tools such as a VHDL compiler and a cycle accurate hardware simulator to support VHBC. The first version of the VHM has been implemented and vindicates the idea of implementing hardware components in VHBC and interpreting it to be viable and feasible.

The main focus of the current work is devoted to an optimized version of the VHM, which will be implemented on the Xilinx Virtex chip. It will be able to provide more functional units and a higher number of registers. In the future we will try to map the VHM design efficiently onto a variety of FPGAs of different vendors by using a pure VHDL description of the VHM similar to the C reference implementation of the Java Virtual Machine (C-Machine).

References

M. Mahadevan, R. M. Bradley, *Journal of Applied Physics*, Vol 79, 1996

M. Budiu, "Application-Specific Hardware: Computing Without CPUs", citeseer.nj.nec.com/497138.html

Y. Ha, P. Schaumont, M. Engels, S. Vernalde, F. Potargent, L. Rijnders, H. de Man, "A Hardware Virtual Machine for the Networked Reconfiguration", In *Proc. of 11th IEEE International Workshop on Rapid System Prototyping (RSP 2000)*, 2000

Y. Ha, S. Vernalde, P. Schaumont, M. Engels, H. De Man, "Building a Virtual Framework for Networked Reconfigurable Hardware and Software Objects", In *Proc. of PDPTA '00*, 2000

S. Goldstein et al., "PipeRench: A Coprocessor for Streaming Multimedia Acceleration", In *Proc. of 24th International Symposium on Computer Architecture*, 1999

Y. Chou, P. Pillai, H. Schmit, and J. P. Shen, "PipeRench Implementation of the Instruction Path Coprocessor", In *Proc. of MICRO '00*, 2000

B. Mei, P. Schaumont, S. Vernalde, "A Hardware-Software Partitioning and Scheduling Algorithm for Dynamically Reconfigurable Embedded Systems"

Y. Ha, B. Mei, P. Schaumont, S. Vernalde, R. Lauwereins, H. De Man, "Development of a Design Framework for Platform-Independent Networked Reconfiguration of Software and Hardware", In *Proc. of FLP*, 2001

R. Kress, *A Fast Reconfigurable ALU for Xputers*, PhD thesis, Universitaet Kaiserslautern, 1996

R. Hartenstein, M. Merz, T. Hoffmann, U. Nageldinger, "Mapping Applications onto reconfigurable KressArrays", In *Proc. of FLP*, 1999

J. Becker, T. Pionteck, C. Habermann, M. Glesner, "Design and Implementation of a Coarse-Grained Dynamically Reconfigurable Hardware Architecture", In *Proc. of Workshop on VLSI (WVLSI)*, 2001

C. Nitsch, U. Kebschull, "The Use of Runtime Configuration Capabilities for Networked Embedded Systems", In *Proc. of DATE'02*, Paris, 2002

S. Guccione, D. Verkest, I. Bolsens, "Design Technology for Networked Reconfigurable FPGA Platforms", In *Proc. of DATE'02*, Paris, 2002

G. De Micheli, "Synthesis and Optimization of Digital Circuits", McGraw-Hill Higher Education, 1994

Chapter 12

A Low Energy Data Management for Multi-Context Reconfigurable Architectures

M. Sanchez-Elez[1], M. Fernandez[1], R. Hermida[1], N. Bagherzadeh[2]

[1] Dpto. Arquitectura de Computadores y Automatica
Universidad Complutense de Madrid
marcos@fis.ucm.es

[2] Dpt. Electrical Engineering and Computer Science University of California, Irvine

Abstract   This paper presents a new technique to improve the efficiency of data scheduling for multi-context reconfigurable architectures targeting multimedia and DSP applications. The main goal of this technique is to diminish application energy consumption. Two levels of on-chip data storage are assumed in the reconfigurable architecture. The Data Scheduler attempts to optimally exploit this storage, by deciding in which on-chip memory the data have to be stored in order to reduce energy consumption. We also show that a suitable data scheduling could decrease the energy required to implement the dynamic reconfiguration of the system.

Keywords: Reconfigurable Computing, Memory Management, Energy Consumption, Multimedia

12.1   Introduction

The emergence of high capacity reconfigurable devices is igniting a revolution in general purpose processors. It is now possible to tailor make and dedicate reconfigurable units to take advantage of application dependent dataflow. The reconfigurable systems combine reconfigurable hardware units with a software programmable processor. They generally have wider applicability than application specific circuits. In addition, they attain a better performance than a general purpose processor for a wide range of computationally intensive applications.

FPGAs [Brown and Rose, 1996] are the most common fine-grained devices used for reconfigurable computing. Dynamic reconfiguration [Tau et al.,

P. Lysaght and W. Rosenstiel (eds.),
New Algorithms, Architectures and Applications for Reconfigurable Computing, 145–155.
© 2005 Springer. Printed in the Netherlands.

1995] has emerged as a particularly attractive technique for minimizing the reconfiguration time, which has a negative effect on FPGA performance. Multi-context architectures are an example of dynamic reconfiguration, which can store a set of different configurations (contexts) in an internal memory; when a new configuration is needed, it is loaded from this internal memory which is faster than reconfiguration from the external memory. Examples of such architectures are MorphoSys [Singh et al., 2000] or FUZION [Meibner et al., 2001].

Multimedia applications are fast becoming one of the dominating workloads for reconfigurable systems. For these applications the system energy is clearly dominated by memory accesses. Most of these applications apply block-partitioning algorithms to the input data; these blocks are relatively small and can easily fit into reasonably sized on-chip memories. Within these blocks there is a significant data reuse as well as data spatial locality [Sohoni et al., 2001]. Therefore, data management at this block level significantly reduces the system energy consumption.

Our previous works [Maestre et al., 1999], [Maestre et al., 2001] and [Sanchez-Elez et al., 2002] discussed scheduling for multi-context architectures, in particular using MorphoSys. The first approach did not deal with energy minimization and different on-chip memory levels, but the method presented in this paper optimizes data storage in different on-chip memory levels, reducing the energy consumption.

Although previous work was done relating to energy optimizations, this kind of data management for reconfigurable systems has not been discussed in detail by other authors. A review of data memory design for embedded systems is discussed in [Panda et al., 2001], but it does not take into account the reconfiguration energy consumption. A method to decrease the power consumption by transforming the initial signal or data processing specifications is suggested in [Wuytack et al., 1994], but it only deals with fine-granularity problems. In [Kaul et al., 1999] a data scheduler for reconfigurable architectures was proposed, though it does not optimize memory management. However, regarding memory organization most efforts were made in cache organization for general purpose computers, see [Kamble and Ghose, 1997] and [Wilson et al., 1995], and not on more custom memory organizations, as needed for example by multimedia and DSP applications.

This paper begins with a brief overview of MorphoSys and its compilation framework. Section 3 describes the problem, and Section 4 analyzes data management in the reconfigurable cells internal memory. A data management in the internal memory to reduce energy consumption is discussed in Section 5. Section 6 analyzes the low energy data management for different on-chip memories. Experimental results are presented in Section 7. Finally we present some conclusions from our research in Section 8.

12.2 Architecture and Framework Overview

This section describes the targeted system M2, the second implementation of MorphoSys. We also present the development framework that integrates the Data Scheduler with other compilation tasks.

MorphoSys (Figure 12.1.a) achieves high performance, compared with other approaches, on several DSP and multimedia algorithms. It consists of an 8×8 Array of Reconfigurable Cells (RC Array). Its functionality and interconnection network are configured through 32-bit context words which are stored in the Context Memory (CM). The data and results are stored in the Frame Buffer (FB) that serves as a data cache (level 1), and it is logically organized into two sets. Data from one set is processed in the current computation, while the other set stores results in the external memory and loads data for the next round of computation. Moreover, each cell has an internal RAM (RC-RAM), level 0 cache, that can be used to store the most frequently accessed data and results. The Data Scheduler developed in this paper reduces data and results transfers, and optimizes their storage to minimize energy consumption. There is also a DMA controller to establish the bridge that connects the external memory to either the FB or the CM. A RISC processor controls MorphoSys operation.

Multimedia and DSP applications typically executed on Morphosys are composed of a group of macro-tasks (kernels), which are characterized by their contexts and their data. Application assembly code and contexts are written in terms of kernels that are available in the kernel library included in the MorphoSys compilation framework (Figure 12.1.b). The Information Extractor generates the information needed by the compilation tasks that follow it, including kernel execution time, data size and the number of contexts for each kernel.

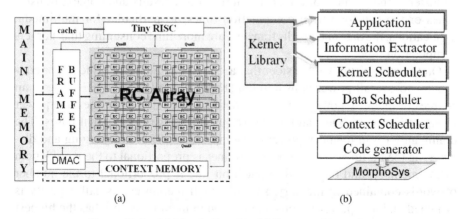

(a)   (b)

Figure 12.1. (a) MorphoSys chip (b) Framework overview.

The Kernel Scheduler explores the design space to generate a possible sequence of kernels that minimizes the execution time. It decides which is the best sequence of kernels and then creates clusters. The term cluster is used here to refer to a set of kernels that are assigned to the same FB set and whose components are consecutively executed. For example, an application is composed of kernels k_1, k_2, k_3, k_4; the kernel scheduler, estimating context and data transfers, could assign k_1 and k_2 to one FB set, and k_3 and k_4 to the other set. This implies the existence of two clusters $c_1 = \{k_1, k_2\}$ and $c_2 = \{k_3, k_4\}$. Whilst the first cluster is being executed using data of one FB set and/or RC-RAMs, meanwhile the contexts and data of the other cluster kernels are transferred to CM and to the other FB set respectively.

The Context and Data Schedulers specify when and how each transfer must be performed to reduce energy consumption. The Code Generator builds the optimized code that is going to be implemented in M2.

12.3   Problem Overview

We propose a new methodology to perform energy efficient data scheduling on a given set of clusters. The problem could be defined as: "*Given an ordered set of clusters with a known size of data and results, and a memory hierarchy, find the data scheduling that minimizes the energy consumption*".

Multimedia applications have a significant data reuse as well as spatial locality at kernel level. The Data Scheduler makes good use of this feature to reduce energy. The data that have a reasonable reuse can be stored in the lowest energy memory in order to minimize energy consumption as explained below.

The MorphoSys execution model allows overlapping of context and data transfers with system computation, but it does not allow simultaneous transfers of data and contexts. Therefore the execution time [Maestre et al., 1999] for a cluster is the greatest of either the sum of data, results and context transfer times or the sum of kernel computation times. In particular for multimedia applications, the data and results transfers are more time consuming than kernels computation time.

On the other hand, we consider that access to the memory is the major contributor to overall power consumption [Kamble and Ghose, 1997]. The main sources of memory power are: the power dissipated in the bit-lines, the power dissipated in word-lines, and the power used to drive the external buses. The dominant term in memory power consumption comes from charging and discharging the bit lines, the number of bit lines is proportional to the word length of the memory, and the length of each bit line is proportional to the number of words contained in memory, so with small memories a small capacity is switched and less power is consumed. Though memory access has the highest power contribution, we also took into account the power dissipated by reconfigurable hardware elements and RISC processor.

The memory access is the major contributor to the overall energy consumption due to power consumption and data access time. The Data Scheduler reduces energy consumption by storing the most frequently used data in the smallest on-chip data memories. In addition, the Data Scheduler allows RF consecutive kernel executions, reducing context transfers by a factor RF (Contexts Reuse Factor), thus, the energy consumption of context memory is also diminished. Moreover, the Data Scheduler reduces data and context transfers, which implies that the energy used to drive the external buses is also reduced.

Although our approach to solve this problem targets one particular reconfigurable system, MorphoSys, this approach is quite general in nature and may be easily applied to other reconfigurable and embedded systems.

12.4 Low Energy RC-RAM Management

Memory design in MorphoSys is hierarchical wherein RC-RAM is the lowest hierarchy level. In the first version of MorphoSys chip (M1) the RC memory was only used to store data for special algorithms because data transfers to/from RC-RAM were very time consuming (7 cycles to transfer one word). These algorithm data were loaded before the application execution and there was no transfer between the FB and the RC-RAM during the execution.

In M2 we have developed new instructions and contexts, which makes one cycle transfers between FB and RC-RAM possible. It also lets the Data Scheduler optimize data storage in on-chip memories. The new instructions and contexts let these transfers to be carried out through DMA. A new Tiny RISC instruction loads DMA with the FB start address, number of words, transfer type (load/store) and RC cell. RC-RAM access is controlled by new contexts, which contain the address register and the transfer type.

On the other hand, MorphoSys speed is limited by the access time to the FB. The Tiny RISC makes the instructions, which do not process FB data, run faster than the dependent ones. As RC-RAMs are within RC, the instructions dependent on these memories are quicker than the instructions in the FB, and they are also smaller than the FB, so they consume less energy. It therefore seems that load data in RC-RAM should reduce energy consumption, but this is not always true. There is an increase in execution time due to the data transfers from the FB to the RC-RAMs having to be done within their cluster computation time, instead of overlapping with previous cluster computation, as occurs in transfers between the external memory and the FB. And there is also an increase in power because data from the external memory must be loaded first into the FB and then into RC-RAM. However, if the data stored in the RC-RAM are read enough times (n_{min}), the decrease in energy consumption from processed data stored in RC-RAMs is larger than the energy wasted in transferring data from FB to RC-RAMs [Sanchez-Elez et al., 2003].

The Data Scheduler attempts to keep the data and results in the RC-RAM until the last kernel that processes these data is executed. EF_{RC} reflects the energy reduced if these data or results are kept in RC-RAM:

$$EF_{RC}(D) = (n_{acc} - n_{min}(d)) \cdot D(u, v)^2 \cdot N^2$$
$$EF_{RC}(R) = (n_{acc} - n_{min}(r)) \cdot R(u, v)^2 \cdot N^2$$

n_{acc}: *number of instructions that read those data or results.* $n_{min}()$: *the minimum number of instructions to improve energy consumption (d: data; r: results).* **D(u,v)**: *the size of input data stored in the RC-RAM for cluster u and processed by cluster v (data shared among clusters).* **R(u,v)**: *the size of results stored in the RC-RAM for cluster u and processed by cluster v (results shared among clusters).*

The Data Scheduler stores data and results following the energy factor until no more data fit into RC-RAM. The data size stored in the RC memories (RCDS) can be obtained by the sum of all sizes of data and results. It takes into account that data have to be stored before the first cluster that uses them is executed. The results are stored while the kernel that produces them is executed and the space occupied by the results or data that are not going to be used again is released and can be used to store new data or results.

$$RCDS = \underset{c \in \{1,\dots,N\}}{MAX} \left\{ \sum_{u=1,v=c}^{c,N} \left(\underset{k \in \{1,\dots,n\}}{MAX} \left[\sum_{i=k}^{n} d_1(u, c) + \sum_{i=1,j=k}^{k,n} r_{ij}(c, v) \right] \right) \right.$$
$$\left. + \sum_{u=1,v=c+1}^{c-1,N} (D(u, v) + R(u, v)) \right\}$$

$d_j(u,c)$: *input data size for kernel k_j of cluster c, these data were stored in RC-RAM for cluster u execution.* $r_{ij}(c,v)$: *intermediate results size of kernel k_j of cluster c which are data for kernel k_j of cluster v, and not for any kernel executed after it.*

The Data Scheduler sorts the data and results to store in RC-RAM according to EF_{RC} . It starts checking that the data with the highest EF_{RC} fit in the RC-RAM. Scheduling continues with data or results with less EF_{RC}. If RCDS > RC-RAM-size for some data or results, these are not stored. The Data Scheduler stores the highest possible amount of data or results that minimizes energy consumption. The Data Scheduler repeats this algorithm for all the 64 RC cells. However, in most cases, this is not necessary because the 64 RC cells have the same execution behavior, though on different data.

12.5 Low Energy FB Management

There are also data shared among clusters or results used as data by some clusters that are not stored in RC-RAMs. The Data Scheduler attempts to keep these data or results in the FB instead of reloading them in each cluster execution, to increase energy reduction. This data transfer reduction between the FB and the external memory was discussed in [Sanchez-Elez et al., 2002]. The main difference between that scheduling and the scheduling proposed here is the RC-RAM memory management. The Data Scheduler finds the external data and intermediate results shared among clusters but not stored in any RC-RAM. It attempts to keep these in the FB instead of reloading them to increase energy reduction. EF_{FB} reflects the energy reduced if these data or results are kept in the FB:

$$EF_{RC}(D) = D(u, v)^2 \cdot (N - 1)^2$$
$$EF_{RC}(R) = R(u, v)^2 \cdot (N + 1)^2$$

The Data Scheduler keeps data and results shared among clusters following the energy factor until no more data fit into the FB. The data size stored in the FB for cluster c (FBDS(c)) can be obtained by the sum of the sizes of all data and results plus the size of the data stored in the FB before transferring them to the RC-RAMs (D_{RC}) and of the results previously stored in the RC-RAMs loaded in the FB before transferring them to the external memory (R_{RC}).

$$FBDS(c) = MAX \left\{ D_{RC}(c) = \sum_{i=1}^{n} d_i, \ \underset{k \in \{1,\dots,n\}}{MAX} \right.$$
$$\times \left[\sum_{i=k}^{n} d_1 = \sum_{i=1}^{k} \left(rout_1 + \sum_{j=k}^{n} r_{i,j} \right) \right] R_{RC}(c) + \sum_{i=1}^{n} rout_1 \right\}$$
$$+ \sum_{u-1, y=c+1}^{c-1, N} (D(u, v) + R(u, v))$$

The Data Scheduler reduces energy because data and results used by cluster kernels are loaded only once even though more than one kernel uses them. Memory write and read operations are executed mainly in the RC-RAMs which consumes less energy than the FB. The Data Scheduler minimizes write and read to/from the external memory due to the fact that it tries to transfer only the input data and the output results of the application, keeping the intermediate results, when possible, in the FB and RC-RAMs.

12.6   Low Energy CM Management

The Data Scheduler also reduces energy consumed by the CM by taking into account multimedia application data characteristics. They are composed of a sequence of kernels that are consecutively executed over a part of the input data, until all data are processed. For example, an algorithm would need to process the total amount of data n times, in this case the contexts of each kernel may be loaded into the CM n times. However, loop fission can be applied sometimes to execute a kernel RF consecutive times before executing the next one (loading kernel data for RF iterations in the FB). In this case kernel contexts are reused because they have to be loaded only n/RF times, reducing context transfers from the external memory and thus minimizing energy consumption. The number of consecutive kernel executions RF (Reuse Factor) is limited by the internal memory sizes. The Data Scheduler finds the maximum RF value taking into account memories and data sizes.

As Morphosys architecture has two on-chip memory levels (FB and RC-RAMs) with different amount of data stored in them, the Data Scheduler can obtain different RF values, RF_{RC} and RF_{FB}, which stand for the reuse factor allowed by the RC-RAMS and by the FB respectively. RF may be the same for all clusters and reconfigurable cells on account of data and results dependencies, the Data Scheduler finds the RF that minimizes the energy consumption taking into account the RF_{RC} and RF_{FB} values, and the data reuse among clusters.

The Data Scheduler finds the RF that minimizes energy as follows:

- If $RF_{FB} \leq RF_{RC}$ then $RF = RF_{FB}$, because all data have to be stored beforehand in FB.

- If $RF_{FB} > RF_{RC}$ the Data Scheduler reduces RF_{FB}, if this means energy reduction, or it increases RF_{RC} reducing data stored in the RC-RAMs if this implies energy reduction. The Data Scheduler does this till $RF_{FB} = RF_{RC} = RF$.

This RF maximizes data reuse, but it does not context reuse. This data and context scheduling minimizes energy but could not be the minimum one. The Data Scheduler increases RF, increasing context reuse, till maximum possible value, and as a consequence reducing data reuse if this implies a reduction in energy consumption. Thus the final RF value obtained by the Data Scheduler minimizes energy consumption, although this value could not maximize data or context reuse (Figure 12.2).

Although the number of memory access could not be minimized, there are other techniques to minimize energy consumption. For example, reducing switching activity on the memory address and data buses produces a decrease in energy consumption. These energy reductions can be brought about by the appropriate reorganization of memory data, thus consecutive memory references exhibit spatial locality. This locality, if correctly exploited, results in

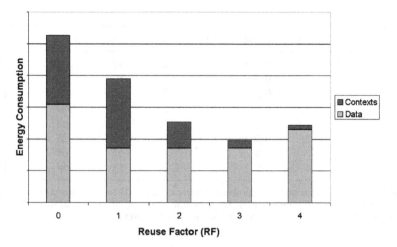

Figure 12.2.   Energy consumption for different RF. RF = 0 without scheduling; RF = 1 maximum data reuse; RF = 4 maximum contexts reuse; RF = 3 Data Scheduler RF.

a power-efficient implementation because, in general, the Hamming distance between nearby addresses is less than that between those that are far apart [Cheng and Pedram, 2001]. During the cycles for which the data-path is idle, all power consumption can then be easily avoided by any power-down strategy. A simple way to achieve this is, for example, the cheap gated-clock approach [Van Oostende and Van Wauve, 1994].

12.7   Experimental Results

In this section we present the experimental results for a group of synthetic and real experiments, in order to demonstrate the quality of the proposed methodology. As a real experiment we have developed a ray-tracing algorithm for MorphoSys [Du et al., 2003]. The algorithm involves projecting rays into the computer's model of the world to determine what color to display at each point in the image.

The data scheduling depends on kernel scheduling, data size and available internal memory. We analyze different kernel schedules for different memory sizes as shown in Table 12.1. We compared the Previous Data scheduler [Sanchez-Elez et al., 2002] and the Data Scheduler proposed in this paper with the Basic Scheduler [Maestre et al., 1999]. E1 stands for the relative energy improvement on the Previous Data Scheduler and E2 stands for the relative energy improvement of the current Data Scheduler. We also did a comparison between the Data scheduler and the Previous Data Scheduler (E1-2).

We tested the same kernel schedules for different memory sizes as shown Table 12.1, A1-A2 or B1-B2 or RT1-RT2. RC-RAM size is always smaller than FB size because RC-RAM is hierarchically the lowest memory. A1 and A2 have

Table 12.1.   Experimental results

	N	n	DS	RAM	FB	RF	E1(%)	E2(%)	E1-2(%)
A1	5	4	1.1	0.125	0.5	2	45	53	18
A2	5	4	1.1	0.06	0.25	1	5	20	16
B1	5	5	2.7	0.25	1	5	60	65	15
B2	5	5	2.7	0.5	1	5	60	69	28
C	4	3	2.5	0.25	1	5	55	68	30
RT1	4	2	2	0.25	1	1	6	26	22
RT2	4	2	2	0.5	2	3	55	68	37
RT3	7	2	4	0.5	2	2	55	58	10

N: total number of clusters; n: maximum number of kernel per cluster; DS: total data size per iteration (input data + intermediate results + final results); RF: reuse context factor; FB: one FB set size in KB; RAM: RC-RAM size in KB; $E1$, $E2$, $E1$-2: Data Schedulers Relative Improvement.

few data stored in RC-RAM since the majority of data are not read many times. The increase in FB size achieves a better reduction in energy consumption by avoiding context transfers.

B1 and B2 have many data stored in RC-RAM. For B1 all the most accessed data cannot be stored in RC-RAM. Therefore an increase in RC-RAM size, as B2 shows, achieves a better result. RT1 and RT2 represent simple image ray-tracing, and for this case an increase in RC-RAM size does not improve energy performance because most of the accessed data fit into the RC-RAM. However an increase in FB and RC-RAMs sizes allows context reuse. Although the increase in memory size is more energy consuming this is worthwhile due to greater data and context reuse. The experiment C stands for an intermediate example between A and B. RT3 stands for a more complicated image ray-tracing, which increases data size and number of kernels. In all cases the Data Scheduler achieves better energy results than the previous version, as E2 and E1-2 show, due to the current Data Scheduler improving of FB, RC-RAMs and CM usage, minimizing energy consumption.

12.8   Conclusions

In this work we have presented a new technique to improve data scheduling for multi-context reconfigurable architectures. It stores the most frequently accessed data in the on-chip memories (FB or RC-RAMs) to minimize data and context transfers, reducing the energy consumption. The Data Scheduler decides which data or results have to be loaded into these internal memories to reduce energy consumption.

The Data Scheduler allows data and results reuse within a cluster, minimizing the memory space required by cluster execution, and enables the reuse of data and results among clusters if the FB has sufficient free space. It chooses the shared data or results to be kept within the FB, allowing further reductions in transfers to/from the external memory.

The Data Scheduler maximizes the available free space in the FB. This enables the number of consecutive iterations (*RF*) to increase and as a consequence, kernel contexts are reused during these iterations, reducing context transfers.

The experimental results demonstrate the effectiveness of this technique in reducing the energy consumption compared with previous data schedulers.

References

Brown, S. and Rose, J. (1996). Architecture of fpgas and clpds: A tutorial. *IEEE Design and Test of Computer*, 13(2):42–57.

Cheng, W. and Pedram, M. (2001). Low power techniques for address encoding and memory allocation. *ASP-DAC Proceedings*, pages 242–250.

Du, H., M., S.-E., and et al, T. N. (2003). Interactive ray tracing on reconfigurable simd morphosys. *ASP-DAC Proceedings*.

Kamble, M. B. and Ghose, K. (1997). Analytical energy dissipation models for low power caches. *Proceedings of the ACM/IEEE International Symposium on Microarchitecture*, pages 184–193.

Kaul, M., R., V., S., G., and I., O. (1999). An automated temporal partitioning and loop fission approach for fpga based reconfigurable synthesis of dsp applications. *Proceedings 36th Design Automation Conference*, pages 616–622.

Maestre, M., F., K., M., F., N., B., and H., S. (2001). Configuration management in multi-context reconfigurable systems for simultaneous performance and power optimizations. *Proceedings of the International Symposium on System Sinthesys*, pages 107–113.

Maestre, M., F., K., N., B., H., S., and M., F. (1999). Kernel scheduling in reconfigurables architectures. *DATE Proceedings*, pages 90–96.

Meibner, M., S., G., W., S., J., P., and D., L. (2001). Parallel volume rendering on a single-chip simd architercture. *IEEE Symposium on Parallel and Large-Data Visuallization and Graphics*.

Panda, P. R., F., C., K., D. N. D. D., E., B., C., K., A, V., and P., K. (2001). Data and memory optimization techniques for embedded systems. *ACM Transactions on Design Automation of Electronic Systems*, 6(2):149–206.

Sanchez-Elez, M., M., F., L., A. M., H., D., N., B., and M., F. (2003). Low energy data management for different on-chip memory levels in multi-context reconfigurable architectures. *DATE Proceedings*, pages 36–41.

Sanchez-Elez, M., M., F., R., M., F., K., R., H., and N., B. (2002). A complete data scheduler for multi-context reconfigurable architectures. *DATE Proceedings*, pages 547–552.

Singh, H., M., L., G., L., F., K., and N., B. (2000). Morphosys: An integrated reconfigurable system for data-parallel and computation-intensive applications. *IEEE Transactions on Computers*, 49(5):465–481.

Sohoni, S., R., M., Z., X., and Y., H. (2001). A study of memory system performance of multimedia applications. *SIGMETRICS Performance 2001*, pages 206–215.

Tau, E., D., C., I., E., J., B., and A., D. H. (1995). A first generation dpga implementation. *Canadian Workshop of Field-Programmable Devices*.

Van Oostende, P. and Van Wauve, G. (1994). Low power design: a gated-clock strategy. *Proceedings of the Low Power Workshop*.

Wilson, P. R., S., J. M., M., N., and D., B. (1995). Dynamic storage application a survey and critical review. *Proceedings of the International Workshop on Memory Management*, pages 1–116.

Wuytack, S., F., C., F., F., L., N., and H, D. M. (1994). Global communication and memory optimizing transformations for low power design. *Proceedings of the International Workshop on Low Power Design IWLPD'94*, pages 203–208.

Chapter 13

Dynamic and Partial Reconfiguration in FPGA SoCs: Requirements Tools and a Case Study

Fernando Moraes, Ney Calazans, Leandro Möller,
Eduardo Brião, Ewerson Carvalho*

* *Pontifícia Universidade Católica do Rio Grande do Sul (PUCRS)*
Av. Ipiranga, 6681—Prédio 30/BLOCO 4—90619—900—Porto Alegre—RS-BRASIL
moraes@inf.pucrs.br

13.1   Introduction

Current technology allows building integrated circuits (ICs) complex enough to contain all major elements of a complete end product, which are accordingly called *Systems-on-Chip* (SoCs) [1]. A SoC usually contains one or more programmable processors, on-chip memory, peripheral devices, and specifically designed complex hardware modules. SoCs are most often implemented as a collection of reusable components named *intellectual property cores* (IP cores or cores). An IP core is a complex hardware module, created for reuse and that fulfills some specific task.

Field Programmable Gate Arrays (FPGAs[1]) have revolutionized the digital systems business after their introduction, almost 20 years ago. The first FPGAs were fairly simple, being able to house a circuit with no more than a few thousand equivalent gates. They were launched in the mid eighties, at the same time when rather complex microprocessors like Intel 80386, Motorola 68020 and the first MIPS were already in the market or were about to be launched. Today, state-of-the-art FPGAs allow accommodating digital systems with more than 10 million equivalent gates in its reconfiguration fabric alone. Such devices can clearly be seen as a platform to implement SoCs. IP cores in such a platform may be implemented using the reconfigurable fabric or hard blocks, available all in the same device. In addition to the reconfigurable fabric, current FPGAs may contain one or a few microprocessors, up to hundreds of integer multipliers and memory blocks, and other specific modules.

P. Lysaght and W. Rosenstiel (eds.),
New Algorithms, Architectures and Applications for Reconfigurable Computing, 157–168.
© 2005 *Springer. Printed in the Netherlands.*

FPGAs are extensively employed today for rapid system prototyping, and in countless finished products. The main FPGA distinguishing feature is the possibility of changing the system hardware structure in the field, a process known as *reconfiguration*. Device reconfiguration enables the incremental development of systems, the correction of design errors after implementation, and the addition of new hardware functions. Systems built with FPGAs can be statically or dynamically reconfigured. A *static reconfigurable system* is one where it is necessary to completely reconfigure it each time any change of the hardware is required. In this case, system execution stops during reconfiguration. A *dynamic reconfigurable system* (DRS) allows that part of the system be modified, while the rest of it continues to operate.

This Chapter focus on the development of DRS design methods targeted to SoCs implemented on FPGAs. The main distinction between conventional digital systems design and the methods addressed here is the possibility of dynamically replacing hardware modules (i.e. IP cores) while the rest of the system is still working.

Current digital system development tools do not present all the features required to adequately design DRSs. Some tools were proposed in the past to allow core parameterization. For example, Luk et al. [2] describe cores using a proprietary language and translate these automatically to VHDL. FPGA vendors offer similar tools to allow core parameterization in their design environment. These are examples of core manipulation early in the design flow, not developed for DRS.

James-Roxby et al. [3] describe a tool called JBitsDiff, based on the JBits API [4]. The user generates a core using a standard design flow, defining its bounding-box with a floorplanning tool. The method presented by Dyer et al. [5] implements the routing between cores using a structure called *virtual socket*, which defines an interface between static and reconfigurable parts of the device. This interface is built from feed-through routed FPGA logic blocks. The virtual socket is manually placed and routed to guarantee correct connection between reconfigurable cores. The PARBIT tool [6] has been developed to transform and restructure bitstreams to implement dynamically loadable hardware modules. PARBIT employs a complete bitstream, a target partial bitstream and parameters provided by the user. PARBIT relies on a modified version of the standard Xilinx FPGA router to achieve predictable routing. The JPG tool [7] is a Java-based partial bitstream generator. JPG, based on the Xilinx JBits API, is able to generate partial bitstreams for Virtex devices based on data extracted from the standard Xilinx CAD physical design flow. Recently, Xilinx proposed a DRS technique using its Modular Design flow to allow a limited form of partial reconfiguration in Virtex devices [8]. To enable the technique, Xilinx made available the *bus macro* component, which must be used to create IP core communication interfaces.

All these works have as common feature the ability for partial bitstream generation. To allow generic reuse of cores by partial bitstream replacement, a more structured approach to provide reconfigurable cores with a communication interface is required. The most relevant contribution of this Chapter is to present and compare two such approaches, one developed by the authors, and another based on the Xilinx DRS technique.

The rest of this work is organized as follows. Section 13.2 presents requirements for FPGA SoC DRSs. Section 13.3 describes a set of tools, proposed to enable the creation of DRSs. Section 13.4 presents a DRS implementation case study, while Section 13.5 draws some conclusions and directions for future work.

13.2   Requirements for FPGA SoC DRSs

DRSs can contain at the same time static IP cores and dynamic IP cores, the latter loaded according to an execution schedule. To achieve partial reconfiguration targeted to core reuse, it is possible to identify six requirements that have to be fulfilled [9].

FPGAs supporting partial and dynamic reconfiguration. Two FPGA vendors currently offer support to partial and dynamic reconfiguration in their devices: Atmel and Xilinx. Atmel produces the AT40K and AT6000 device families. Atmel devices are not considered further, since their small capacity makes them inadequate to support FPGA SoCs. Xilinx offers the Virtex, Virtex-II, and Virtex-II Pro families. The atomic reconfigurable unit in Xilinx devices is a frame, a vertical section extending from top to bottom of the device. An IP core is physically implemented in such devices as a set of consecutive frames.

Tools to determine the IP core position and shape. Placement restrictions must be attached to reconfigurable IPs during DRS design to define their size, shape and position. The Xilinx Floorplanner is an example of a tool that can be used to generate such restrictions.

Support for partial bitstream generation and download. The total and partial reconfiguration protocols are distinct. Thus, special features must be present in tools used to partially reconfigure FPGAs.

Communication interface between cores. A communication structure among IP cores is necessary to allow both, the exchange of data and control signals during normal operation, and insulation during reconfiguration. In a conventional SoC design, standard on-chip bus architectures, such as AMBA and CoreConnect may be used to provide IP core communication. However, these architectures do not provide features needed in DRSs like core insulation during reconfiguration. To enable dynamic insertion or removal of cores into an operating FPGA, a communication interface to connect and insulate cores at different moments must be available. This interface should provide several functions such as arbitration, communication between modules, input/output pins virtualization and configuration control.

SoC input/output pins virtualization. Most SoCs present a fixed interface to the external world, even if internally they are reconfigurable. This is done to restrict the hardware reconfiguration complexity to the intra-chip environment. Thus, an internally reconfigurable SoC needs to provide I/O capability to every reconfigurable IP it may possibly lodge. I/O pins virtualization is a generic feature allowing IP cores to exchange data only with their communication interface, while extra-chip communication is handled by the SoC fixed interfaces.

Dynamic reconfiguration control. A method to control or execute the dynamic reconfiguration process must be available. This can be implemented in software, in hardware or using a mix of both.

The first three requirements can be satisfied with existing devices and tools. To the authors' knowledge, works fulfilling the last three requirements do not exist ready for use in commercial devices. In addition, related academic works do not address communication structures.

13.3   Tools for DRS

This Section presents two approaches to achieve DRS. The first was developed by the authors, and is based on a tool called *CoreUnifier*, while the second is based on a DRS technique proposed by Xilinx.

CoreUnifier

The method based on this tool relies on the usage of a shared global bus, which is the communication media to interconnect IP cores. It is possible to dynamically insert and/or remove cores using this communication structure.

The method is divided into three main phases. First, the designer generates a complete bitstream for each core, using any conventional CAD tool flow. Each core must have virtual pins, implemented with tristate buffers connected to a shared bus. In addition, area constraints must be generated for each IP core, using e.g. floorplanning tools. The second step is the use of the CoreUnifier tool itself. This tool accepts as input a set of complete bitstreams, one of which, called *master bitstream*, comprises static modules responsible for control and communication. Each of the other complete bitstreams contains one or more reconfigurable cores of the final system. The CoreUnifier extracts from each of these bitstreams the area segment corresponding to the specific IP core description and merges it with the appropriate area segment of the master bitstream. This produces every partial bitstream corresponding to one reconfigurable IP core. The third phase is the DRS operation that starts when the user downloads

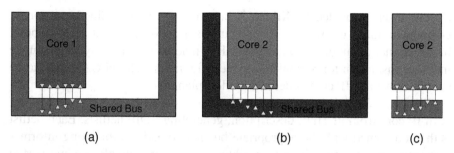

Figure 13.1. CoreUnifier usage: (a)-(b) complete bitstreams showing the communication interface and a reconfigurable core; (c) the partial bitstream generated by the CoreUnifier tool.

the master bitstream. Dynamic reconfiguration then proceeds according to some scheduling scheme. A simple tool, called *BitScheduler* was also developed as a proof-of-concept to help validating the method.

Figure 13.1 illustrates the CoreUnifier usage. Figure 13.1(a) and Figure 13.1(b) shows two distinct cores connected to the shared bus. Assume that Figure 13.1(a) is the master bitstream. It is important to emphasize that the physical implementation of the area below **Core 1** and **Core 2** may be different, since each one was obtained with different synthesis steps using non-deterministic placement and routing algorithms. Figure 13.1(c) shows the replacement of **Core 1** by **Core 2** into the master bitstream and the partial bitstream generation. The CoreUnifier replaces only the area above the shared bus, keeping the previous physical implementation. This was the challenge to implement this tool, since FPGA vendors do not distribute complete internal information regarding their devices physical structure.

Xilinx Modular Design based DRS Technique

The DRS technique proposed by Xilinx is based on the Modular Design flow, a method originally proposed to support team-based system development [8]. The Modular Design flow is divided in three phases: *Initial Budgeting*, *Active Module Implementation* and *Final Assembly*. The first phase corresponds to design partitioning, while the second is related to individual IP cores implementation, and the third regards IP cores integration. The DRS technique based on Modular Design assumes that each reconfigurable or fixed module is implemented as a contiguous set of FPGA logic block columns. Also, the technique requires that intra-chip dynamic module communication be strictly point-to-point.

The *Initial Budgeting* phase starts with the generation of a top module described in a hardware description language (HDL). This top module instantiates each IP core of the SoC, but do not contain any IP core internal description. All reconfigurable modules are connected among them or to fixed modules with the bus macro provided by Xilinx. A bus macro is a pre-defined, pre-routed component, made up by eight tristate buffers, providing a 4-bit point-to-point bi-directional communication channel. Bus macros are one way to guarantee interface insulation for regions under reconfiguration. The FPGA region to be occupied by each IP core is defined by floorplanning.

The *Active Modules Implementation* phase starts with the logic synthesis of each IP core from its HDL description, generating EDIF netlists. Each netlist is then combined with the appropriate bus macro and floorplanning information from the previous phase to produce each reconfigurable module partial bitstream.

The *Final Assembly* phase is where system level information about IP core interfaces, external pins and floorplanning are combined with the fixed and re-configurable IP cores physical design information, to produce the initial complete bitstream. This bitstream must then be used to configure the FPGA, after what partial bitstreams obtained in the second phase can be used to perform partial and dynamic SoC reconfiguration.

To allow partial automation of the Modular Design flow, a tool called MD-Launcher has been proposed and implemented by the authors [10] to decrease the amount of manual errors made during implementation. Using the tool, the user is expected to create the top module and the module files only, using these as input to the tool, which then automatically performs all phases of the Modular Design flow.

The capital difference between the CoreUnifier based and the Xilinx DRS techniques is the communication interface definition. The CoreUnifier is targeted to core reuse, since a shared bus is employed to provide data exchange between cores, while the Xilinx DRS technique provides only point-to-point communication using bus macro components. Another difference is that the CoreUnifier-based technique needs that physical design tools be enhanced before it can be fully deployed for automated execution. The current available version of these tools do not allow the required finer control of the positioning and wiring of elements composing the interface between any reconfigurable IP core and the rest of the device.

13.4   A DRS Case Study: Design and Experimental Results

Possibly, the most intuitive form of DRS is one where the instruction set of a processor is dynamically incremented through the use of reconfigurable hardware. Such components are generically known as *reconfigurable processors*. The first choice that has to be made when implementing a reconfigurable processor regards the degree of coupling between the basic processor structure (datapath and control unit) and the reconfigurable units. *Tightly coupled*

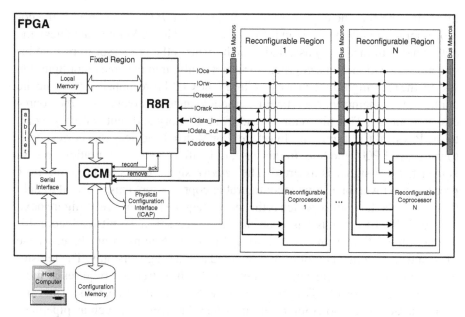

Figure 13.2. General structure of the R8R system implementation.

reconfigurable processors rely upon the dynamic modification of the datapath and/or the control unit to provide extension to the native Instruction Set Architecture (ISA). *Loosely coupled* reconfigurable processors usually provide a standard interface to reconfigurable units and employ specialized instructions to allow communication between these and the processor units. Examples of tightly coupled approaches are the classical DISC [11] and the Altera Nios [12]. The latter approach can be exemplified by the Garp architecture [13] and the case study described herein (R8R).

Figure 13.2 displays the conceptual organization of the R8R system implementation. The system is composed by three main modules: *(i)* a host computer, providing an interface to the system user; *(ii)* a configuration memory, containing all partial bitstream files used during the system execution; *(iii)* the FPGA, containing fixed and reconfigurable regions of the R8R system.

The fixed part in the FPGA is a complete computational system, comprising the R8R processor, its local memory containing instructions and data, a system bus controlled by an arbiter and peripherals (serial interface and CCM). The serial interface peripheral provide capabilities for communication with the host computer (an RS232-like serial interface). The Configuration Control Manager (CCM) peripheral is a specialized hardware, designed for acting as a slave of the R8R processor or the host computer. The host computer typically fills or alters the configuration memory before system execution starts.

The normal operation of the CCM module is to wait for the R8R processor to produce coprocessor reconfiguration requests using the *reconf* signal, while informing the specific coprocessor identifier in the *IOaddress* lines. If the co-processor is already in place (an information stored in CCM internal tables), the *ack* signal is immediately asserted, which releases the processor to resume instruction execution. If the coprocessor is not configured in some reconfigurable region, the reconfiguration process is fired. The CCM is responsible to locate the configuration memory area where lies the adequate bitstream from the coprocessor identifier. This bitstream is read from memory and sent, word by word, to the Physical Configuration Interface. For Virtex-II devices this corresponds to the ICAP FPGA module. In this case, only after the reconfiguration process is over the *ack* signal is asserted. The *remove* signal exists to allow the R8R processor to invalidate some reconfigurable coprocessor. This is useful to help the CCM to better choose which is the most adequate region to reconfigure next.

The R8R processor is based on the R8 processor, a 16-bit load-store 40-instruction RISC-like processor [14]. R8 is a von Neumann machine, whose logic and arithmetic operations are executed among registers only. The register bank contains 16 general-purpose registers. Each instruction requires at most 4 clock cycles to execute. The processor is a RISC-like machine, but still missing some characteristics so common in most RISC processors, such as pipelines. The original R8 processor was transformed into the R8R processor by the addition of five new instructions, intended to give support to the use of partially reconfigurable coprocessors. The added instructions are defined and described in Table 13.1. The R8R processor was wrapped to provide communication with the *(i)* local memory; *(ii)* the system bus; *(iii)* the CCM; *(iv)* the reconfigurable

Table 13.1. Instructions added to the R8 processor in order to produce the R8R processor.

Reconfigurable instruction	Semantics description
SELR *address*	Selects the coprocessor identified by *address* for communication with the processor, using the *reconf* signal. If the coprocessor is not currently loaded into the FPGA, the CCM automatically reconfigures some region of the FPGA with it.
INITR *address*	Resets the coprocessor specified by *address*, using the *IOreset* signal. The coprocessor must have been previously configured.
WRR *RS1 RS2*	Sends the data stored in *RS1* and *RS2* to the coprocessor selected by the last *selr* instruction. *RS1* can be used for passing a command while *RS2* passes data. The coprocessor must have been previously configured.
RDR *RS1 RT2*	Sends the data stored in *RS1* to the coprocessor (typically a command or an address). Next, reads data from the coprocessor, storing it into the *RT2* register. The coprocessor must have been previously configured.
DISR *address*	Informs the CCM, using the *remove* signal, that the coprocessor specified by *address* can be removed if needed.

regions. The interface to the reconfigurable regions comprises a set of signals connected to bus macros. Partial bitstreams for the R8R system are generated using the previously described Xilinx DRS design flow.

The processor-coprocessor communication protocol is based on read/write operations. Write operations are used by the processor to send data to some coprocessor. A write operation uses *IOce, IOrw, IOdata_out* and *IOaddress* signals, broadcasting these to all reconfigurable regions. The reconfigurable region containing the address specified by the *IOaddress* signal gets the data. The read operation works similarly, with the difference that data flows from a coprocessor to the processor. The *IOreset* is asserted by the processor to initialize one of the coprocessors. The communication among processor and coprocessors is achieved by using bus macros. In this communication scheme the processor is the system master, starting all read/write operations. The next version of this system will include the use of interrupt controlled communication, which will allow to implement non blocking operations and coprocessor-initiated communication.

The above described system has been completely prototyped and is operational in two versions, with one and two reconfigurable regions, respectively. A V2MB1000 prototyping platform from Insight-Memec was used. This platform contains a million-gate XC2V1000 Xilinx FPGA, memory and I/O resources. For validation purposes, the whole system was partitioned in two blocks: one comprising the CCM and the Physical Configuration Interface modules, and another with the rest of the system. Each block was validated separately. The merging of the blocks is currently under way.

In order to evaluate the relative performance of the implemented reconfigurable system, a set of experiments was conducted. For each experiment, a functionality was implemented in both software and hardware, as a coprocessor. Each implementation was executed in a system version with one reconfigurable region, and their execution times were compared. The execution time for the hardware version in each experiment includes not only the coprocessor execution time, but also the time to load the coprocessor into the FPGA by reconfiguration. Figure 13.3 shows the comparison between the number of operations executed and the time to execute hardware and in software versions of three 16/32-bit arithmetic nodules: multiplication, division and square root. Note that for each module, the execution time grows linearly but with different slopes for software and hardware implementations. Also, the reconfiguration time adds a fixed latency (10 ms) to the hardware implementations. The break even point for each functionality determines when a given hardware implementation starts to be advantageous with regard to a plain software implementation, based on the number of times this hardware is employed before it is reconfigured. The multiplier, division and square root coprocessors are advantageous from 750, 260 and 200 executions, respectively. This simple experiment highlights how in

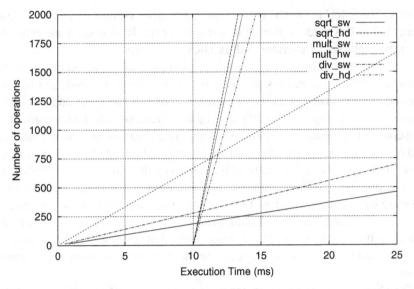

Figure 13.3. Execution time versus the number of operations for three arithmetic modules, multiplication, division and square root, implemented in hardware (_hw suffix) and software (_sw suffix). The hardware reconfiguration latency, 10ms, is dependent on the size of the reconfigurable region partial bitstream and on the performance of the CCM module. This graph was plotted for a CCM working at 24 MHz frequency and for reconfigurable bitstreams of 46 Kbytes, corresponding to a reconfigurable region with an area of roughly one tenth of the employed million-gate device. With more design effort it is estimated that this latency can be reduced by at least one order of magnitude, for the same bitstream size [10].

practice it is possible to take advantage of DRSs, achieving performance gains, flexibility and system area savings.

13.5   Conclusions

A DRS potentially reduces the final system cost, since the user can employ smaller FPGAs, downloading partial bitstreams on demand. In addition, partial reconfiguration makes possible for systems to benefit from a *virtual hardware* approach, in the same manner that e. g. computers benefit from the use of virtual memory.

It should be clear after the above discussion that partial and dynamic reconfigurable FPGA SoCs are viable to be implemented. However, there are several open issues that must be addressed in order to allow the technology to become mainstream.

First, the lack of tools to deal with the differences between a static SoC design flow and a DRS SoC design flow is acute and crucial. It is acute, since available tools are only incremental additions to static flows and most of them address only lower level tasks automation. It is crucial since, unless an automated flow exists to support DRS, no industrial adoption of DRS can be expected. Second, just one vendor really gives device support to FPGA SoCs. Even these devices are not ideal for a seamless DRS design flow. For example, there is a relatively small number of tristate components available in Virtex FPGAs, limiting the development of complex communication interfaces for DRS.

However, it is already possible to devise some trends arising in the DRS horizon. There is a trend to incrementally add hardware and software support for DRS. For example, Xilinx recently introduced the JBits 3.0 API to manipulate partial bitstreams in its Virtex-II family of FPGAs. In the hardware side, an internal module in Virtex-II FPGAs provides access to the configuration interface from inside the device, a feature absent in previous FPGA families. As enhancements are made available, a boost in the DRS use can be expected, providing positive feedback to work in the area.

The DRS application area is another crucial point. Unless sound applications are developed to show real advantage of a DRS design solution over conventional solutions, DRS will remain no more than an academic exercise on an interesting subject area.

Notes

1. Any reference to FPGAs here concerns RAM based FPGAs, where configuration and reconfiguration are synonyms.

References

[1] R. Bergamaschi; et al. Automating the Design of SOCs using Cores. *IEEE Design & Test of Computers*, vol 18(5), 2001. pp. 32–45.

[2] W. Luk; S. McKeever. Pebble: a Language for Parametrised and Reconfigurable Hardware Design. In: *Field Programmable Logic and Applications (FPL)*, pp. 9–18, 1998.

[3] P. James-Roxby; S. Guccione. Automated Extraction of Run-time Parameterisable Cores from Programmable Device Configurations. In: *Field-Programmable Custom Computing Machines (FCCMt'00)*, pp. 153–161, 2000.

[4] Xilinx, Inc. The JBits 2.8 SDK for Virtex, 2001.

[5] M. Dyer; C. Plessl; M. Platzner. Partially Reconfigurable Cores for Xilinx Virtex. In: *Field Programmable Logic and Applications (FPL)*, pp. 292–301, 2002.

[6] E. Horta; J. Lockwood; S. Kofuji. Using PARBIT to implement Partial Run-time Reconfigurable Systems. In: *Field Programmable Logic and Applications (FPL)*, pp. 182–191, 2002.

[7] A. Raghavan; P. Sutton. JPG—a Partial Bitstream Generation Tool to Support Partial Reconfiguration in Virtex FPGAs. In: *International Parallel and Distributed Processing Symposium (IPDPS)*, pp. 155–160, 2002.

[8] Xilinx, Inc. Two Flows for Partial Reconfiguration: Module Based or Difference Based. Application Note XAPP290, Version 1.1. http://www.xilinx.com/xapp/xapp290.pdf, November, 2003.

[9] D. Mesquita; F. Moraes; J. Palma; L. Mller; N. Calazans. Remote and Partial Reconfiguration of FPGAs: Tools and Trends. In: *10th Reconfigurable Architectures Workshop (RAW)*, 2003.

[10] E. Brião. Partial and Dynamic Reconfiguration for Intellectual Property Cores. *MSc Dissertation, PUCRS, Faculty of Informatics*. 110 p. January, 2004. (In Portuguese)

[11] M. Wirthlin and B. Hutchings. A Dynamic Instruction Set Computer. In: *Field-Programmable Custom Computing Machines (FCCM'95)*, pp. 99–107, 1995.

[12] Altera Corp. Nios Embedded Processor System Development. http://www. altera.com/products/ip/ processors/nios/nios-index.html, 2003.

[13] T. Callahan; J. Hauser, J. Wawrzynek. The Garp Architecture and C Compiler. *IEEE Computer*, vol 33(4), pp. 62–69, 2000.

[14] F. Moraes; N. Calazans. R8 Processor—Architecture and Organization Specification and Design Guidelines. http://www.inf.pucrs.br/~ gaph/Projects/R8/public/R8_arq_spec_eng.pdf, 2003.

Applications

Chapter 14

Design Flow for a Reconfigurable Processor
Implementation of a Turbo-decoder

Alberto La Rosa, Luciano Lavagno, and Claudio Passerone

Politecnico di Torino
Dipartimento di Elettronica
corso Duca degli Abruzzi 24
10129 Torino—Italy
{larosa,lavagno,passerone}@polito.it

Abstract   This paper describes an approach to hardware/software design space exploration
for reconfigurable processors. The existing compiler tool-chain, because of the
user-definable instructions, needs to be extended in order to offer developers an
easy way to explore the design space. Such extension often is not easy to use
for developers who have only a software background, and are not familiar with
reconfigurable architecture details or hardware design. Our approach differs from
others because it is based on a simple extension of the standard programming
model well known to software developers. We illustrate it by using a real case
study, a software implementation of a UMTS turbo-decoder that achieves a $11\times$
speed-up of the whole application, by exploiting user-defined instructions on the
dynamically reconfigurable portion of the datapath.

Keywords: FPGA, Reconfigurable Hardware, Design Flow, Software Development, Turbo-
decoding

14.1   Introduction

Reconfigurable computing is emerging as a promising means to tackle the
ever-rising cost of design and masks for Application-Specific Integrated Circuits. Adding a reconfigurable portion to a Commercial Off The Shelf IC enables it to potentially support a much broader range of applications than a
more traditional ASIC or microcontroller or DSP would. Moreover, *run-time
reconfigurability* even allows one to adapt the hardware to changing needs and
evolving standards, thus sharing the advantages of embedded software, but still

P. Lysaght and W. Rosenstiel (eds.),
New Algorithms, Architectures and Applications for Reconfigurable Computing, 171–181.
© 2005 *Springer. Printed in the Netherlands.*

achieve higher performance and lower power than a traditional processor, thus partially sharing the advantages of custom hardware.

A key problem with dynamically reconfigurable hardware is the inherent difficulty of programming it, since neither the traditional synthesis, placement and routing-based hardware design flow, nor the traditional compile, execute, debug software design flow directly support it. In a way, dynamic reconfiguration can be viewed as a hybrid between:

- software, where the CPU is "reconfigured" at every instruction execution, while memory is abundant but has limited access bandwidth, and

- hardware, where reconfiguration occurs seldom and very partially, while memory is relatively scarce but has potentially a very high access bandwidth.

Several approaches, some of which are summarized in the next section, have been proposed to tackle this problem, both at the architecture and at the CAD level.

In this case study, we consider the XiRisc architecture [3], in which a reconfigurable unit called pGA has been added as one of the function units of a RISC processor. This approach has the huge advantage, over the co-processor-based alternatives, that it can exploit, as discussed in the rest of this paper, a standard software development flow. The pGA can be managed directly by the compiler, so today the designer just needs to tag portions of the source code that must be implemented as single pGA-based instructions. The disadvantage of the approach is that the reconfigurable hardware accesses its main data memory via the normal processor pipeline (load and store instructions), thus it partially suffers from the traditional processor bottleneck. Without a direct connection to main memory, however, we avoid to face memory consistency problems [6, 2], thus simplifying the software programming model.

In this paper we show that, by efficiently organizing the layout of data in memory, and by exploiting the dynamic reconfiguration capabilities of the processor, one can obtain an impressive speed-up (better than $11\times$) for a very compute-intensive task, like turbo-decoding.

The goal thus is to bridge the huge distance between software and hardware, currently at several orders of magnitude, in terms of cost, speed and power. The result is a piece of hardware that is reconfigured every 100 or so clock cycles (at least in the application described in this paper; other trade-offs are possible for applications with a more static control pattern), and within that interval is devoted to a single computation task. Ideally, we would like to bring a traditional, hardware-bound task, that of decoding complex robust codes for wireless communication, within the "software" computation domain.

14.2 Related Work

Coupling a general purpose processor with a reconfigurable unit generally requires to embed them on the same chip, otherwise communication bandwidth limitations may severely impact the performance of the entire system. There are several ways of designing the interaction between them, but they can be grouped into two main categories: (i) the reconfigurable array is a function unit of the processor pipeline; (ii) the reconfigurable array is a co-processor communicating with the main processor. In both cases, the unit can be reprogrammed on the fly during execution to accommodate new kernels, if the available space does not permit to have all of them at the same time. Often, a cache of configurations is maintained close to the array to speed up this process.

In the first category we find designs such as PRISC [8], Chimaera [5, 10] and ConCISe [7]. In these examples the reconfigurable array is limited to combinational logic only, and data is read and written directly from/to the register file of the processor.

The second category includes the GARP processor [2]. Since the reconfigurable array is external, there is an overhead due to the explicit communication using dedicated instructions that are needed to move data to and from the array. But because of the explicit communication, the control hardware is less complicated with respect to the other case, since almost no interlock is needed to avoid hazards. If the number of clock cycles per execution of the array is relatively high, then the overhead of communication may be considered negligible.

In all the cited examples a modified C compiler is used to program the processor and to extract candidate kernels to be downloaded on the array. In particular, the GARP compiler [2] and the related Nimble compiler [9] identify loops and find a pipelined implementation, using profiling-based techniques borrowed from VLIW compilers and other software optimizations.

Tensilica offers a configurable processor, called Xtensa, where new instructions can be easily added at design time within the pipeline. Selection of the new instructions is performed manually by using a simulator and a profiler. When the Xtensa processor is synthesized, a dedicated development toolset is also generated that supports the newly added instruction as function intrinsics.

A Tensilica processor is specialized for a given algorithm at fabrication time. On the other hand, a reconfigurable processor can be customized directly by the software designer. Of course, this added flexibility has a cost: a pGA-based implementation of the specialized functional unit has area, delay and power cost that is about 10 times larger than an ASIC implementation, such as Xtensa.

The architecture we are considering falls in the first of the above categories, but it allows sequential logic within the array, thus permitting loops to be executed entirely in hardware. The software development flow, on the other hand, is similar to the Tensilica one, in that we rely on manual identification of the extracted computational kernels. However, we do not require re-generation of the complete tool chain whenever a new instruction is added.

14.3 Design Flow for the Reconfigurable Processor

We are targeting a reconfigurable processor, called XiRisc, that is being developed by the University of Bologna and is described in [3]. Here we will briefly outline the main characteristics, that are useful to understand the rest of the paper.

The processor is based on a 32-bit RISC (DLX) reference architecture. However, both the instructions set and the architecture have been extended to support:

- Dedicated hardware logic to perform multiply-accumulate calculation and end-of-loop condition verification.

- A double data-path, with concurrent VLIW execution of two instructions. The added data-path is limited to only arithmetic and logic instructions, and cannot directly access the external memory subsystem.

- A dynamically reconfigurable array called pGA (for pico Gate Array) to implement in hardware special kernels and instructions. The pGA can be dynamically reconfigured at run-time and can store up to 4 configurations that can be switched every clock cycle by the instruction decoder. Each pGA row is composed of 16 4-input, 2-output Look-Up-Tables and 32 register bits and includes a fast carry chain. Different configuration share the state of the registers inside the array.

Comparing reconfigurable processors with traditional ones, software developers have to adopt an extended programming model that includes reconfigurable units besides their standard instruction set architecture.

Such kind of extensions often require developers to provide architecture-specific information and to execute user-defined instructions directly through assembler code or compiler intrinsics (pseudo-function calls), thus lowering the abstraction level provided by generic programming languages like C or C++. As consequence, the main drawback of such approaches is that software developers need to master reconfigurable processor details and hardware description languages before being productive in writing software for such kind of computer architectures.

The main goal of compiler tools is that to hide architecture details, by providing a standardized programming model (variables, arithmetic and logical operators, control constructs and functions and procedure calls) that permits code reuse among different computer architectures.

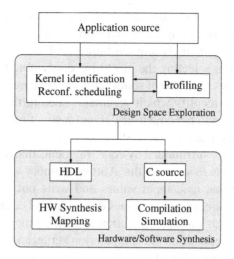

Figure 14.1. The proposed design flow.

Our approach to reconfigurable computing tries to extend the standard programming model to include easy and transparent use of the reconfigurable instruction set.

The design flow, shown in Figure 14.1, is divided in two main phases:

- *Design space exploration*: here software developers analyze existing source code (often written without specific knowledge of the target processor), and identify groups of statements that may be implemented as user-defined instructions on the pGA. Each new instruction is characterized by cost figures (latency, power, number of required CLBs) and a software functional model, later used in ISS simulation for performance estimation and functional verification. In this way several hardware/software partitions can be explored and compared.

- *Software and hardware synthesis*: this phase represents the final link to the target architecture, when the binary code and the bit-stream for the pGA are generated through software and hardware compiler tools. Starting from the functional model of each user-defined instruction, a netlist of gates is synthesized and mapped on pGA CLBs. In this step real cost figures are computed. Then the software compilation chain can be used again, this time with real instruction latency information to optimize instruction scheduling and register allocation.

The design exploration methodology that we propose is primarily aimed at optimizing performance in term of speed, making use of a limited number of pGA cells. The reduction in number of executed instructions and memory accesses also has a very beneficial effect on energy consumption.

The first step consists of identifying which statements within the application, if implemented in hardware, may improve performance. We use common profiling tools (like gprof), analyzing the timing and the number of executions of each function. After selecting the most time consuming functions, a further profiling step annotates each line of source code with the number of cycles spent executing it. Thus we identify which statements need further investigation and are possible candidates to become user-defined instructions.

In most cases, the chosen group of statements is a computation kernel inside a loop. Such kernels are surrounded by code fragments that access data memory to read input values and to store results. Among memory access statements, we distinguished those that read input values and write output data, from those which read and write intermediate values during kernel computation. The first set of accesses depends on the application and on its memory data structure, while the second set depends on the computation kernel and its implementation.

At this point one can perform two partially orthogonal kinds of design space exploration:

- optimize memory data structures in order to minimize memory accesses for input, output and intermediate values [4].

- implement in hardware the entire computation kernel (or part of it) using the pGA.

After extracting the control-data flow graph (CDFG) of each identified kernel, we decompose it into non overlapping subgraphs that can be implemented in hardware as a set of pGA instructions. In our target architecture the pGA unit can read up to four input operands from the register file and write up to two results to it, with a maximum of four register accesses per instruction due to instruction encoding constraints. Thus we looked for subgraphs having up to four input edges and up to two output edges, for a total of at most four.

By exploring various splittings of the CDFG, different possible implementations of the computation kernel on our reconfigurable processor can be tried. Then a trade-off analysis between cost and performance will help in selecting the optimal implementation.

In order to evaluate cost and performance, in the case of hardware implementation, we characterize each subgraph according to the estimated latency and number of required CLBs. These numbers can be derived manually if the developer has enough knowledge of digital hardware design, or through an automated estimator available in our software development chain.

After partitioning, we recompile the source code with our compiler, which schedules instructions so as to best exploit both the standard data-paths and the pGA, by using the latency assigned to each pGA instruction. Then we use an instruction set simulator to analyze system performance. Once the optimal implementation is selected, a finer analysis of performance and functionality is possible through RTL simulation of the entire system (core and pGA). We first synthesize hardware starting from an HDL translation of each CFDG subgraph and then we place and route the pGA. Information extracted from reports will help in refining user-defined instruction latencies and costs, that were previously only estimated.

14.4   Design Tools for the Reconfigurable Processor

We implemented a prototype of the design flow described above to support the reconfigurable processor. It starts from a *fully* C initial specification, where sections that must be moved to the pGA are manually annotated, and automatically generates the assembler code, the simulation model, and a hardware model useful for instruction latency and datapath cost estimation. A key characteristic is that it supports compilation and simulation of software including user-definable instructions, without the need to recompile the tool chain every time a new instruction is added.

The design flow is based on the gcc tool chain, which was modified to meet our requirements. We re-targeted the compiler by changing the machine description files found in the gcc distribution, to describe the extensions to the DLX architecture and ISA, such as the second data-path and the reconfigurable unit.

We had to modify both the scheduling and the generation of the object code of the gcc assembler to support the XiRisc processor architecture. The scheduler needs to handle the second data-path, and some static scheduling to avoid the parallel issue of two instructions that read and write the same register is required. We also modified both the assembler instruction mnemonics and their binary encodings to add two classes of instructions: one for the DSP instructions and another for the pGA instructions.

We finally modified gdb for simulation and debugging: the simulation models for DSP instructions were permanently added to the simulator, since they are a fixed part of the architecture. On the other hand, pGA instructions depend on the application, and thus cannot be statically linked.

To simulate a pGA instruction, we use a C model of its behavior, which is called at simulation time when the instruction is encountered. The latency of the instruction is also specified by the model, and it is used when computing the performance. A pGA instruction is identified in the source code using a pragma directive. Figure 14.2-(a) shows an example of annotated source code and Figure 14.2-(b) shows the code automatically generated by the modified tool chain. If macro PGA is defined, the generated code is compiled for the target processor. If the macro is not defined, the generated code is compiled for simulation; in this case, function _shift_add is used as the simulation model.

```
int bar (int a, int b) {
    int c;
#pragma pga shift_add 0x12 5 1 2 c a b
    c = (a << 2) + b;
#pragma end
    return c + a;
}
```

```
int bar (int a, int b) {
    int c;
#if defined(PGA)
    asm("fpga5 0x12, %0,%1,%2": "=r"(c): "r"(a),
        "r"(b));
#else
    asm("tofpga %1,%2,$0": :"r"(a), "r"(b));
    asm("jal _shift_add");
    asm("fmfpga %0,$0,$0": "=r"(c));
#endif
    return c + a;
}
#if !defined(PGA)
void _shift_add () {
    int c, a, b;
    asm("move %0,$2;move %1,$3": "=r"(a),
        "=r"(b));
    c = (a << 2) + b;
    /* delay by 5 cycles */
    asm("move $2,%0; li $4,5": : "r"(c));
}
#endif
```

(a)   (b)

Figure 14.2.

Simulating the VLIW architecture is fairly straightforward: since the target architecture is almost fully bypassed, and since the assembler already prevents possible structural hazards by inserting nop instructions, it is sufficient to simulate instructions sequentially even though they are executed in parallel on the processor.

Performance information is computed during simulation. We assume that the processor never stalls for non-pGA instructions, since all such hazards are resolved at compilation time. Therefore, it is sufficient to increment the cycle count at each instruction that is simulated, with two notable exceptions:

1. pGA instructions should increment the clock cycle count by a possibly variable amount (which is actually computed by the simulation code).

2. Two parallel VLIW instructions are executed in the same clock cycle. Therefore, only one of them should increment the clock cycle count.

We also developed a simple profiler which takes a trace as input and generates the total number of clock cycles of the simulation run, the total number of clock cycles spent for each different instruction (opcode), and the total number of clock cycles spent in each line of the source code.

14.5   Case Study: Turbo Decoding

We consider the specification of a turbo-decoder that follows the 3GPP recommendations for UMTS cellular phone systems [1]. The turbo-decoder algorithm has many advantages over other encoding/decoding mechanisms when the transmission channels has very low signal to noise ratios. However, it has a high computational complexity, which has limited its application until recently, when enough computational power has become available.

A very important aspect of the optimization was a re-organization of the memory layout, and a conversion of floating point values to fixed point for the target processor, which does not have a floating point unit. By appropriately pairing short ints into a single memory word, the memory access bandwidth, one of the bottlenecks of the DLX-based architecture that we are using, can be doubled. We also re-arranged the code so that accesses to arrays in main memory was minimized. Since these are standard transformations, that have little to do with the reconfigurable nature of the processor, we refer the interested reader to [4] for more details. *All comparisons between reconfigurable and non-reconfigurable processor below are given based on the memory-optimized version of the source code, for both processors.* Thus they consider only the effects of the pGA, not of any other optimization.

The decoder implements an iterative algorithm, repeated until convergence. Each iteration consists of several components, among which the most computationally intensive ones are: (1) trellis metrics computation (γ); (2) forward probabilities computation (α); (3) backward probabilities computation (β); (4) maximum likelihood ratio computation (LLR).

We analyzed the CDFG of the above kernels, and extracted candidate pGA instructions that were then mapped on the reconfigurable array. Some of the kernels, such as determining forward and backward probabilities, share similar configurations (in this case a butterfly computation) that differ only due to the order in which memory data are read and written. Therefore we implemented two pGA instructions, one for the butterfly, and another for reordering. In the case of the LLR computation, we also explored different approximations of the max* operator, which is used to reduce multiplications to additions in the logarithmic domain.

In most cases, a single pGA instruction was not sufficient to cover the entire kernel. Thus we extensively exploited both the ability to store four different configurations in the pGA, thus supporting four different user-defined instructions at any given point in time, and the fact that those instructions share the values of the registers in every row of the pGA. This is mostly due to the limitation in the number of inputs and outputs that can be concurrently read and written in the same clock cycle.

Table 14.1. Experimental results on 40-bit messages

Step	Speed-up	Exec. cycles	Saved cycles
Original	1×	177834	–
Gamma	1.02×	173706	4128
LLR	1.83×	96913	80921
Butterfly	1.53×	115816	62018
Reorder	1.10×	161826	15972
Final	11.73×	15157	162677

We performed several simulations of the turbo-decoder, with different objectives. The first ones were aimed at analyzing the algorithm to find the best implementation and to drive the selection of dynamic instructions. In particular, we wanted to:

- determine the best performance for different implementations of the max* operator;

- determine the minimum bit-widths for input, output and intermediate data, that still yield satisfactory results;

- determine the best threshold to stop the iterative process and declare convergence.

After selecting and mapping dynamic instructions, we performed several simulations to determine the performance of the turbo-decoder, comparing it to a similarly optimized pure software implementation on the same architecture, but without the pGA. Table 14.1 shows the experimental results that we obtained when different combinations of dynamic instructions are used. Column **Speed-up** indicates the improvement in performance *for the entire algorithm*, including also portions that were not implemented with dynamic instructions. **Exec. cycles** shows the total number of clock cycles needed to complete the decoding, and **Saved cycles** is the difference with respect to the original pure software implementation.

We first simulated the algorithm without any dynamic instructions, and then re-simulated it using only one type of dynamic instruction at a time. Computation of the butterfly operator and of the maximum likelihood ratio proved to be the most expensive functions, and thus benefitted most from hardware acceleration. When all dynamic instructions are used at the same time, the speed-up is a remarkable 11×. Although the two instructions of metrics computation and reordering show modest speed-ups when compared with the software-only implementation, their effect on the final result is very important. In fact, while their contribution to the total latency of the algorithm is low at the beginning, they become the bottlenecks when the other more complex operations are moved into hardware. For instance, if reordering were not implemented as a dynamic instruction, the total number of clock cycles would be around 31.000, corresponding to a speed-up of only $5.7\times$.

14.6 Conclusions

In this paper we showed how a reconfigurable processor can be used to dramatically speed up the execution of a highly data and control intensive task, the decoding of turbo-codes. We used a design flow that enables a software designer, who is mostly unaware of hardware design subtleties, to quickly assess the costs and gains due to executing selected pieces of C code as single instructions on a pGA-based reconfigurable DLX functional unit.

The result is a factor of $11\times$ speed-up on the whole decoding algorithm. It was achieved, while simultaneously optimizing the memory layout, with only about 1 month of work by a software designer, who did not have any previous exposure to hardware design, synthesis, or reconfigurable computing.

References

[1] 3GPP. Technical specification group radio access network; multiplexing and channel coding. Technical Specification TS 25.212 v5.1.0, 2002.

[2] T. Callahan and J. Wawrzynek. Adapting software pipelining for reconfigurable computing. In *Proceedings of the International Conference on Compilers, Architecture, and Synthesis for Embedded Systems (CASES)*, 2000.

[3] F. Campi, R. Canegallo, and R. Guerrieri. IP-reusable 32-bit VLIW Risc core. In *Proceedings of the European Solid-State Circuits Conference*, September 2001.

[4] F. Catthoor, S. Wuytack, E. De Greef, F. Balasa, L. Nachtergaele, and A. Vandecapelle, editors. *Custom Memory Management Methodology: Exploration of Memory Organisation for Embedded Multimedia System Design*. Kluwer Academic Publishers, 1998.

[5] S. Hauck, T. Fry, M. Hosler, and J. Kao. The Chimaera reconfigurable functional unit. In *Proceedings of the IEEE Symposium on FPGAs for Custom Computing Machines*, April 1997.

[6] J. Jacob and P. Chow. Memory Interfacing and Instruction Specification for Reconfigurable Processors. In *Proc. ACM Intl. Symp. on FPGAs*, 1999.

[7] B. Kastrup, A. Bink, and J. Hoogerbrugge. ConCISe: A compiler-driven CPLD-based instruction set accelerator. In *Proceedings of the Seventh Annual IEEE Symposium on Field-Programmable Custom Computing Machines*, April 1999.

[8] R. Razdan and M. Smith. A high-performance microarchitecture with hardware-programmable functional units. In *Proceedings of the 27th Annual International Symposium on Microarchitecture*, November 1994.

[9] L. Yanbing, T. Callahan, R. Harr, and U. Kurkure. Hardware-software co-design of embedded reconfigurable architectures. In *Proceedings of the Design Automation Conference*, June 2000.

[10] Z.A. Ye, N. Shenoy, and P. Banerjee. A C compiler for a processor with a reconfigurable functional unit. In *Proceedings of the ACM/SIGDA International Symposium on Field Programmable Gate Arrays*, February 2000.

Chapter 15

IPsec-Protected Transport of HDTV over IP*

Peter Bellows[1], Jaroslav Flidr[1], Ladan Gharai[1],
Colin Perkins[1], Pawel Chodowiec, and Kris Gaj[2]

[1] *USC Information Sciences Institute, 3811 N. Fairfax Dr. #200,*
Arlington VA 22203, USA;
{pbellows|jflidr|ladan|csp}@isi.edu

[2] *Dept. of Electrical and Computer Engineering, George Mason University,*
4400 University Drive, Fairfax VA 22030, USA;
{pchodow1|kgaj}@gmu.edu

Abstract   Bandwidth-intensive applications compete directly with the operating system's
network stack for CPU cycles. This is particularly true when the stack performs
security protocols such as IPsec; the additional load of complex cryptographic
transforms overwhelms modern CPUs when data rates exceed 100 Mbps. This
paper describes a network-processing accelerator which overcomes these bottlenecks by offloading packet processing and cryptographic transforms to an intelligent interface card. The system achieves sustained 1 Gbps host-to-host bandwidth
of encrypted IPsec traffic on commodity CPUs and networks. It appears to the
application developer as a normal network interface, because the hardware acceleration is transparent to the user. The system is highly programmable and can
support a variety of offload functions. A sample application is described, wherein
production-quality HDTV is transported over IP at nearly 900 Mbps, fully secured
using IPsec with AES encryption.

15.1   Introduction

As available network bandwidth scales faster than CPU power [Calvin,
2001], the overhead of network protocol processing is becoming increasingly
dominant. This means that high-bandwidth applications receive diminishing

*This work is supported by the DARPA Information Technology Office (ITO) as part of the Next Generation
Internet program under Grants F30602-00-1-0541 and MDA972-99-C-0022, and by the National Science
Foundation under grant 0230738.

P. Lysaght and W. Rosenstiel (eds.),
New Algorithms, Architectures and Applications for Reconfigurable Computing, 183–194.
© 2005 *Springer. Printed in the Netherlands.*

marginal returns from increases in network performance. The problem is greatly compounded when security is added to the protocol stack. For example, the IP security protocol (IPsec) [ips, 2003] requires complex cryptographic transforms which overwhelm modern CPUs. IPsec benchmarks on current CPUs show maximum throughput of 40–90 Mbps, depending on the encryption used [Fre, 2002]. With 1 Gbps networks now standard and 10 Gbps networks well on their way, the sequential CPU clearly cannot keep up with the load of protocol and security processing. By contrast, application-specific parallel computers such as FPGAs are much better suited to cryptography and other streaming operations. This naturally leads us to consider using dedicated hardware to offload network processing (especially cryptography), so more CPU cycles can be dedicated to the applications which use the data.

This paper describes a prototype of such an offload system, known as "GRIP" (Gigabit-Rate IPsec). The system is a network-processing accelerator card based on Xilinx Virtex FPGAs. GRIP integrates seamlessly into a standard Linux implementation of the TCP/IP/IPsec protocols. It provides full-duplex gigabit-rate acceleration of a variety of operations such as AES, 3DES, SHA-1, SHA-512, and application-specific kernels. To the application developer, all acceleration is completely transparent, and GRIP appears as just another network interface. The hardware is very open and programmable, and can offload processing from various levels of the network stack, while still requiring only a single transfer across the PCI bus. This paper focuses primarily on our efforts to offload the complex cryptographic transforms of IPsec, which, when utilized, are the dominant performance bottleneck of the stack.

As a demonstration of the power of hardware offloading, we have successfully transmitted an encrypted stream of live, production-quality HDTV across a commodity IP network. Video is captured in an HDTV frame-grabber at 890 Mbps, packetized and sent AES-encrypted across the network via a GRIP card. A GRIP card on a receiving machine decrypts the incoming stream, and the video frames are displayed on an HDTV monitor. All video processing is done on the GRIP-enabled machines. In other words, the offloading of the cryptographic transforms frees enough CPU time for substantial video processing with no packet loss on ordinary CPUs (1.3 GHz Pentium III).

This paper describes the hardware, device driver and operating system issues for building the GRIP system and HDTV testbed. We analyze the processing bottlenecks in the accelerated system, and propose enhancements to both the hardware and protocol layers to take the system to the next levels of performance (10 Gbps and beyond).

15.2   GRIP System Architecture

The overall GRIP system is shown in Figure 15.1. It is a combination of an accelerated network interface card, a high-performance device driver, and

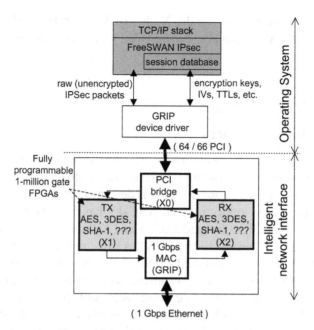

Figure 15.1. GRIP system architecture.

special interactions with the operating system. The interface card is the SLAAC-1V FPGA coprocessor board [Schott et al., 2003] combined with a custom Gigabit Ethernet mezzanine card. The card has a total of four FPGAs which are programmed with network processing functions as follows. One device (X0) acts as a dedicated packet mover/PCI interface, while another (GRIP) provides the interface to the Gigabit Ethernet chipset and common offload functions such as IP checksumming. The remaining two devices (X1 and X2) act as independent transmit and receive processing pipelines, and are fully programmable with any acceleration function. For the HDTV demonstration, X1 and X2 are programmed with AES-128 encryption cores.

The GRIP card interfaces with a normal network stack. The device driver indicates its offload capabilities to the stack, based on the modules that are loaded into X1 and X2. For example in the HDTV application, the driver tells the IPsec layer that accelerated AES encryption is available. This causes IPsec to defer the complex cryptographic transforms to the hardware, passing raw IP/IPsec packets down to the driver with all the appropriate header information but no encryption. The GRIP driver looks up security parameters (key, IV, algorithm, etc.) for the corresponding IPsec session, and prefixes these parameters to each packet before handing it off to the hardware. The X0 device fetches the packet across the PCI bus and passes it to the transmit pipeline (X1). X1 analyzes the packet headers and security prefix, encrypting or providing other security services as specified by the driver. The packet, now completed, is sent to the Ethernet interface on the daughter card. The receive pipeline is just the inverse, passing through the X2 FPGA for decryption. Bottlenecks in other layers of the stack can also be offloaded with this "deferred processing" approach.

15.3 GRIP Hardware

15.3.1 Basic platform

The GRIP hardware platform provides an open, extensible development environment for experimenting with 1 Gbps hardware offload functions. It is based on the SLAAC-1V FPGA board, which was designed for use in a variety of military signal processing applications. SLAAC-1V has three user-programmable Xilinx Virtex 1000 FPGAs (named X0, X1 and X2) connected by separate 72-bit systolic and shared busses. Each FPGA has an estimated 1 million equivalent programmable gates with 32 embedded SRAM banks, and is capable of clock speeds of up to 150 MHz. The FPGAs are connected to 10 independent banks of 1 MB ZBT SRAM, which are independently accessible by the host through passive bus switches. SLAAC-1V also has an on-board flash/SRAM cache for storing FPGA bitstreams, allowing for rapid run-time reconfiguration of the devices. For the GRIP project, we have added a custom Gigabit Ethernet mezzanine card to SLAAC-1V. It has a Vitesse 8840 Media Access Controller (MAC), and a Xilinx Virtex 300 FPGA which interfaces to the X0 chip through a 72-bit connector. The Virtex 300 uses 1 MB of external ZBT SRAM for packet buffering, and performs common offload functions such as filtering and checksumming.

The GRIP platform defines a standard partitioning for packet processing, as described in section 15.2. As described, the X0 and GRIP FPGAs provide a static framework that manages basic packet movement, including the MAC and PCI interfaces. The X0 FPGA contains a packet switch for shuttling packets back and forth between the other FPGAs on the card, and uses a 2-bit framing protocol ("start-of-frame"/"end-of-frame") to ensure robust synchronization of the data streams. By default, SLAAC-1V has a high-performance DMA engine for mastering the PCI bus. However, PCI transfers for a network interface are small compared to those required for the signal processing applications targeted by SLAAC-1V. Therefore the DMA engine was specially tuned with key features needed for high-rate network-oriented traffic, such as dynamic load balancing, 255-deep scatter-gather tables, programmable interrupt mitigation, and support for misaligned transfers.

With this static framework in place, the X1 and X2 FPGAs are free to be programmed with any packet-processing function desired. To interoperate with the static framework, a packet-processing function simply needs to incorporate a common I/O module and adhere to the 2-bit framing protocol. SLAAC-1V's ZBT SRAMs are not required by the GRIP infrastructure, leaving them free to be used by packet-processing modules. Note that this partitioning scheme is not ideal in terms of resource utilization—less than half of the circuit resources in X0 and GRIP are currently used. This scheme was chosen because it provides a clean and easily programmable platform for network research. The GRIP hardware platform is further documented in [Bellows et al., 2002].

15.3.2   X1/X2 IPsec Accelerator Cores

A number of packet-processing cores have been developed on the SLAAC-1V/GRIP platform, including AES (Rijndael), 3DES, SHA-1, SHA-512, SNORT-based traffic analysis, rules-based packet filtering (firewall), and intrusion detection [Chodowiec et al., 2001, Grembowski et al., 2002, Hutchings et al., 2002]. For the secure HDTV application, X1 and X2 were loaded with 1 Gb/s AES encryption cores. We chose a space-efficient AES design, which uses a single-stage iterative datapath with inner-round pipelining. The cores support all defined key sizes (128, 192 and 256-bit) and operate in either CBC or counter mode. Because of the non-cyclic nature of counter mode, the counter-mode circuit can maintain maximum throughput for a single stream of data, whereas the CBC-mode circuit requires two interleaved streams for full throughput. For this reason, counter mode was used in the demonstration system.

The AES cores are encapsulated by state machines that read each packet header and any tags prefixed by the device driver, and separate the headers from the payload to be encrypted/decrypted. The details of our AES designs are given in [Chodowiec et al., 2001]. We present FPGA implementation results for the GRIP system in section 15.7.

15.4   Integrating GRIP with the Operating System

The integration of the hardware presented in the section 15.3 is a fairly complex task because, unlike ordinary network cards or crypto-accelerators, GRIP offers potential services to all layers of the OSI architecture. Offloading IPsec—the main focus of current GRIP research—is particularly problematic, as the IPsec stack does not properly reside in any one layer; rather, it could be described as a link-layer component "wrapped" in the IP stack (Figure 15.2). Care must be taken to provide a continuation of services even though parts of higher layers have been offloaded.

For this study we used FreeSWAN, a standard implementation of IPsec for Linux [fre, 2003]. FreeSWAN consists of two main parts: KLIPS and Pluto. KLIPS (KerneL IP Security) contains the Linux kernel patches that implement the IPsec protocols for encryption and authentication. Pluto negotiates the Security Association (SA) parameters for IPsec-protected sockets using the ISAKMP protocol. Figure 15.2 illustrates the integration of FreeSWAN into

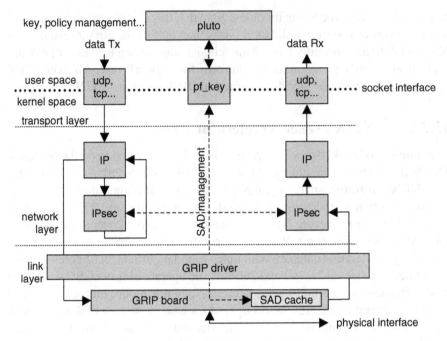

Figure 15.2. The IPsec (FreeSWAN) stack in the kernel architecture.

the system architecture. When Pluto establishes a new SA, it is sent to the IPsec stack via the pf_key socket, where it is stored in the Security Association Database (SAD). At this point, the secure channel is open and ready to go. Any time a packet is sent to an IPsec-protected socket, the IPsec transmit function finds the appropriate SA in the database based on the target IP address, and performs the required cryptographic transforms. After this processing, the packet is returned to the IP layer, which passes it on to the physical interface. The receive mode is the inverse but somewhat less complex. When there are recursive IPsec tunnels, the above process can repeat many times.

In order to accommodate GRIP acceleration we made three modifications to the FreeSWAN IPsec implementation. First, we modified Pluto so that AES Counter mode is the preferred encryption algorithm for negotiating new SA's. Second, we added the ability for the Security Association Database to echo new SA parameters to the appropriate physical interface, using the private space of the corresponding driver. When the GRIP driver receives new SA parameters, it caches encryption keys and other information on the GRIP card for use by the accelerator circuits. Finally, the IPsec transmit and receive functions were slightly modified to support AES counter mode. Any packet associated with an AES SA gets processed as usual—IPsec headers inserted, initialization vectors generated, etc. The only difference is that the packet is passed back to the stack without encrypting the payload. The GRIP driver recognizes these partially-processed packets and tags them with a special prefix that instructs the card to perform the encryption.

15.5 Example Application: Encrypted Transport of HDTV over IP

15.5.1 Background

To demonstrate the performance of the GRIP system, we chose a demanding real-time multimedia application: transport of High Definition Television (HDTV) over IP. Studios and production houses need to transport uncompressed video through various cycles of production, avoiding the artifacts that are an inevitable result of multiple compression cycles. Local transport of uncompressed HDTV between equipment is typically done with the SMPTE-292M standard format for universal exchange [Society of Motion Picture and Television Engineers, 1998]. When production facilities are distributed, the SMPTE-292M signal is typically transported across dedicated fiber connections between sites, but a more economical alternative is desirable. We consider the use of IP networks for this purpose.

15.5.2 Design and Implementation

In previous work [Perkins et al., 2002] we have implemented a system that delivers HDTV over IP networks. The Real-time Transport Protocol (RTP) [Schulzrinne et al., 1996] was chosen as the delivery service. RTP provides media framing, timing recovery and loss detection, to compensate for the inherent unreliability of UDP transport. HDTV capture and playout was via DVS HD-station cards [HDs, 2003], which are connected via SMPTE-292M links to an HDTV camera on the transmitter and an HDTV plasma display on the receiver. These cards were inserted into workstations with standard Gigabit Ethernet cards and dual PCI busses (to reduce contention with the capture/display cards).

The transmitter captures the video data, fragments it to match the network MTU, and adds RTP protocol headers. The native data rate of the video capture is slightly above that of Gigabit Ethernet, so the video capture hardware is programmed to perform color sub-sampling from 10 to 8 bits per component, for a video rate of 890 Mbps. The receiver code takes packets from the network, reassembles video frames, corrects for the effects of network timing jitter, conceals lost packets, and renders the video.

Each video frame is 1.8 million octets in size. To fit within the 9000 octet Gigabit Ethernet MTU, frames are fragmented into approximately 200 RTP packets for transmission. The high packet rates are such that a naive implementation can saturate the memory bandwidth; accordingly, a key design goal is to avoid data copies. We implement scatter send and receive (implemented using the recvfrom() system call with MSG_PEEK to read the RTP header, followed by a second call to recvfrom() to read the data) to eliminate data marshalling overheads. Throughput of the system is limited by the interrupt processing and DMA overheads. We observe a linear increase in throughput as the MTU is increased, and require larger than normal MTU to successfully support the full data rate. It is clear that the system operates close to the limit, and that adding IPsec encryption is not feasible without hardware offload.

This HDTV-over-IP testbed served as the benchmark for GRIP performance. The goal was to add line-rate AES cryptography to the system, without any degradation in video throughput and with no special optimizations to the application. Since the GRIP card appears to the system as a standard Ethernet card, it was possible to transparently substitute a GRIP card in place of the normal Ethernet, and run the HDTV application unmodified. The performance results of this test are given in section 15.7.

15.6   Related Work

Two common commercial implementations of cryptographic acceleration are *VPN gateways* and *crypto-accelerators*. The former approach is limited in that it only provides security between LANs with matching hardware (datalink layer security), not end-to-end (network layer) security. The host-based crypto-accelerator reduces the CPU overhead by offloading cryptography, but overwhelms the PCI bus at high data rates. GRIP differs from these approaches in that it is a reprogrammable, full system solution, integrating accelerator hardware into the core operation of the TCP/IP network stack.

A number of other efforts have demonstrated the usefulness of dedicated network processing for accelerating protocol processing or distributed algorithms. Examples of these efforts include HARP [Mummert et al., 1996], Typhoon [Reinhardt et al., 1994], RWCP's GigaE PM project [Sumimoto et al., 1999], and EMP [Shivam et al., 2001]. These efforts rely on embedded processor(s) which do not have sufficient processing power for full-rate offload of complex operations such as AES, and are primarily primarily focused on unidirectional traffic. Other research efforts have integrated FPGAs onto NICs for specific applications such as routing [Lockwood et al., 1997], ATM firewall [McHenry et al., 1997], and distributed FFT [Underwood et al., 2002]. These systems accelerate end applications instead of the network stack, and often lacked the processing power of the GRIP card.

15.7   Results

15.7.1   System Performance

The HDTV demonstration system was built as described in section 15.5.2, with symmetric multiprocessor (SMP) Dell PowerEdge 2500 servers (2x1.3 GHz) running Linux 2.4.18, substituting a GRIP card in place of the standard Ethernet. The full, 890 Mbps HDTV stream was sent with GRIP-accelerated AES encryption and no compression. In addition, we tested for maximum encrypted bandwidth using iperf [ipe, 2003]. Application and operating system bottlenecks were analyzed by running precision profiling tools for 120 second intervals on both the transmitter and receiver. Transmitter and receiver profiling results are comparable, therefore only the transmitter results are presented for brevity. The profiling results are given in Figure 15.3.

The HDTV application achieved full-rate transmission with no packets dropped. Even though the CPU was clearly not overloaded (idle time > 60%!), stress tests such as running other applications simultaneously showed that the system was at the limits of its capabilities. Comparing the SMP and UP cases under iperf, we can see that the only change (after taking into account the 2X factor of available CPU time under SMP) is the amount of idle time. Yet in essence, the performance of the system was unchanged.

To explain these observations, we consider system memory bandwidth. We measured the peak main memory bandwidth of the test system to be 8 Gbps with standard benchmarking tools. This means that in order to sustain gigabit network traffic, each packet can be transferred at most 8 times to/from main memory. We estimate that standard packet-processing will require three memory copies per packet: from the video driver's buffer to the hdtv application buffer, from the application buffer to the network stack, and a copy within the stack to allow IPsec headers to be inserted. The large size of the video buffer inhibits effective caching of the first copy and the read-access of the second copy. This means these copies consume 3 Gbps of main memory bandwidth for 1 Gbps network streams. Three more main memory transfers occur in writing the video frame from the capture card to the system buffer, flushing ready-to-transmit packets from the cache, and reading packets from memory to the GRIP card. In all, we estimate that a 1 Gbps network stream consumes 6 Gbps of main memory bandwidth on this system. Considering that other

Library/Function	bandwidth	idle	kernel	IPsec	grip driver	appplication	libc
HDTV-SMP	893 Mbps	62%	28%	4%	3%	<1%	< 1%
iperf-SMP	989 Mbps	47%	35%	4%	4%	2%	8%
iperf-UP	989 Mbps	0%	70%	9%	4%	3%	12%

Figure 15.3.   Transmitter profiling results running the HTDV and iperf applications, showing percentage of CPU time spent in various functions.

Design	CLB Util.	BRAM Util	Pred. Perf. (MHz / Gbps)	Measured Perf. (MHz / Gbps)
X0	47%	30%	PCI: 35 / 2.24	33 / 2.11
			I/O: 54 / 1.73	33 / 1.06
X1 / X2 (AES)	17%	65%	CORE: 90 / 1.06	90 / 1.06
			I/O: 47 / 1.50	33 / 1.06
GRIP	35%	43%	41 / 1.33	33 / 1.06
Other modules:				
3DES	31%	0%	77 / 1.57	83 / 1.69
SHA-1	16%	0%	64 / 1.00	75 / 1.14
SHA-512	23%	6%	50 / 0.62	56 / 0.67

Figure 15.4. Summary of FPGA performance and utilization on Virtex 1000 FPGAs.

system processes are also executing and consuming bandwidth, and that the random nature of network streams likely reduces memory efficiency from the ideal peak performance, we conclude that main memory is indeed the system bottleneck.

15.7.2 Evaluating Hardware Implementations

Results from FPGA circuit implementations are shown in Figure 15.4. As shown in the figure, the static packet-processing infrastructure easily achieves 1 Gbps throughput. Only the AES and SHA cores have low timing margins. Note that there are more than enough resources on SLAAC-1V to combine both AES encryption and a secure hash function at gigabit speeds. Also note that the target technology, the Virtex FPGA family, is five years old; much higher performance could be realized with today's technology.

15.8 Conclusions and Future Work

Network performance is currently doubling every eight months [Calvin, 2001]. Modern CPUs, advancing at the relatively sluggish pace of Moore's Law, are fully consumed by full-rate data at modern line speeds, and completely overwhelmed by full-rate cryptography. This disparity between network bandwidth and CPU power will only worsen as these trends continue. In this paper we have proposed an accelerator architecture that attempts to resolve these bottlenecks now and can scale to higher performance in the future. The unique contributions of this work are not the individual processing modules themselves; for example, 1 Gbps AES encryption has been demonstrated by many others. Rather, we believe the key result is the full system approach to integrating accelerator hardware directly to the network stack itself. The GRIP card is capable of completing packet processing for multiple layers of the stack. This gives a highly efficient coupling to the operating system, with only one pass across the system bus per packet. We have demonstrated this system running at full 1 Gbps line speed with end-to-end encryption on commodity PCs. This provides significant performance improvements over existing implementations of end-to-end IPsec security.

As demonstrated by the HDTV system, this technology is very applicable to signal processing and rich multimedia applications. It could be applied to several new domains of secure applications, such as immersive media (e.g. the collaborative virtual operating room), commercial media distribution, distributed military signal processing, or basic VPNs for high-bandwidth networks.

We would like to investigate other general-purpose offload capabilities on the current platform. A 1 Gbps secure hash core could easily be added to the processing pipelines to give accelerated encryption and authentication simultaneously. More functions could be combined by using the rapid reconfiguration capabilities of SLAAC-1V to switch between a large number of accelerator functions on-demand. Packet sizes obviously make a big difference—larger packets mean less-frequent interrupts. The GRIP system could leverage this by incorporating TCP/IP fragmentation and reassembly, such that PCI bus transfers are larger than what is supported by the physical medium. Finally, several application-specific kernels could be made specifically for accelerating the HDTV system, such as RTP processing and video codecs.

Our results suggest that as we look towards the future and consider ways to scale this technology to multi-gigabit speeds, we must address the limitations of system memory bandwidth. At these speeds, CPU-level caches are of limited use because of the large and random nature of the data streams. While chipset technology improvements help by increasing available bandwidth, performance can also greatly improve by reducing the number of memory copies in the network stack. For a system such as GRIP, three significant improvements are readily available. The first and most beneficial is a direct DMA transfer between the grabber/display card and the GRIP board. The second is the elimination of the extra copy induced by IPsec, by modifying the kernel's network buffer allocation function so that the IPsec headers are accommodated. The third approach is to implement the zero-copy socket interface.

FPGA technology is already capable of multi-gigabit network acceleration. 10-Gbps AES counter mode implementations are straightforward using loop-unrolling [Jarvinen et al., 2003]. Cyclic transforms such as AES CBC mode and SHA will require more aggressive techniques such as more inner-round pipelining, interleaving of data streams, or even multiple units in parallel. We believe that 10 Gbps end-to-end security is possible with emerging commodity system bus (e.g. PCI Express), CPU, and network technologies, using the offload techniques discussed.

194

References

[Fre, 2002] (2002). *IPsec Performance Benchmarking*, *http://www.freeswan.org/freeswan_trees/ freeswan-1.99/doc/performance.html*. FreeS/WAN.

[HDs, 2003] (2003). *http://www.dvs.de/*. DVS Digital Video Systems.

[fre, 2003] (2003). *http://www.freeswan.org/*. FreeS/Wan.

[ips, 2003] (2003). *Latest RFCs and Internet Drafts for IPsec, http://ietf.org/html.charters/ipsec-charter.html*. IP Security Protocol (IPsec) Charter.

[ipe, 2003] (2003). *Network performance measuring tool, http://dast.nlanr.net/Projects/Iperf/*. National Laboratory for Applied Network Research.

Bellows, P., Flidr, J., Lehman, T., Schott, B., and Underwood, K. D. (2002). GRIP: A reconfigurable architecture for host-based gigabit-rate packet processing. In *Proc. of the IEEE Symposium on Field-Programmable Custom Computing Machines*, Napa Valley, CA.

Calvin, J. (2001). Digital convergence. In *Proceedings of the Workshop on New Visions ofr Large-Scale Networks: Research and Applications*, Vienna, Virginia.

Chodowiec, P., Gaj, K., Bellows, P., and Schott, B. (2001). Experimental testing of the gigabit IPsec-compliant implementations of Rijndael and Triple-DES using SLAAC-1V FPGA accelerator board. In *Proc. of the 4th Int'l Information Security Conf.*, Malaga, Spain.

Grembowski, T., Lien, R., Gaj, K., Nguyen, N., Bellows, P., Flidr, J., Lehman, T., and Schott, B. (2002). Comparative analysis of the hardware implementations of hash functions SHA-1 and SHA-512. In *Proc. of the 5th Int'l Information Security Conf.*, Sao Paulo, Brazil.

Hutchings, B. L., Franklin, R., and Carver, D. (2002). Assisting network intrusion detection with reconfigurable hardware. In *Proc. of the IEEE Symposium on Field-Programmable Custom Computing Machines*, Napa Valley, CA.

Jarvinen, K., Tommiska, M., and Skytta, J. (2003). Fully pipelined memoryless 17.8 Gbps AES-128 encryptor. In *Eleventh ACM International Symposium on Field- Programmable Gate Arrays (FPGA 2003)*, Monterey, California.

Lockwood, J. W., Turner, J. S., and Taylor, D. E. (1997). Field programmable port extender (FPX) for distributed routing and queueing. In *Proc. of the ACM International Symposium on Field Programmable Gate Arrays*, pages 30–39, Napa Valley, CA.

McHenry, J. T., Dowd, P. W., Pellegrino, F. A., Carrozzi, T. M., and Cocks, W. B. (1997). An FPGA-based coprocessor for ATM firewalls. In *Proc. of the IEEE Symposium on FPGAs for Custom Computing Machines*, pages 30–39, Napa Valley, CA.

Mummert, T., Kosak, C., Steenkiste, P., and Fisher, A. (1996). Fine grain parallel communication on general purpose LANs. In *In Proceedings of 1996 International Conference on Supercomputing (ICS96)*, pages 341–349, Philadelphia, PA, USA.

Perkins, C. S., Gharai, L., Lehman, T., and Mankin, A. (2002). Experiments with delivery of HDTV over IP networks. Proc. of the 12th International Packet Video Workshop.

Reinhardt, S. K., Larus, J. R., and Wood, D. A. (1994). Tempest and typhoon: User-level shared memory. In *International Conference on Computer Architecture*, Chicago, Illinois, USA.

Schott, B., Bellows, P., French, M., and Parker, R. (2003). Applications of adaptive computing systems for signal processing challenges. In *Proceedings of the Asia South Pacific Design Automation Conference*, Kitakyushu, Japan.

Schulzrinne, H., Casner, S., Frederick, R., and Jacobson, V. (1996). RTP: A transport protocol for real-time applications. RFC 1889.

Shivam, P., Wyckoff, P., and Panda, D. (2001). EMP: Zero-copy OS-bypass NIC-driven Gigabit Ethernet message passing. In *Proc. of the 2001 Conference on Supercomputing*.

Society of Motion Picture and Television Engineers (1998). Bit-serial digital interface for high-definition television systems. SMPTE-292M.

Sumimoto, S., Tezuka, H., Hori, A., Harada, H., Takahashi, T., and Ishikawa, Y. (1999). The design and evaluation of high performance communication using a Gigabit Ethernet. In *International Conference on Supercomputing*, Rhodes, Greece.

Underwood, K. D., Sass, R. R., and Ligon, W. B. (2002). Analysis of a prototype intelligent network interface. *Concurrency and Computing: Practice and Experience*.

Chapter 16

Fast, Large-scale String Match for a 10 Gbps FPGA-based NIDS

Ioannis Sourdis[1] and Dionisios Pnevmatikatos[2]

[1] *Microprocessor and Hardware Laboratory, Electronic and Computer Engineering Department, Technical University of Crete, Chania, GR 73 100, Greece*
{sourdis,pnevmati}@mhl.tuc.gr

[2] *Institute of Computer Science (ICS), Foundation for Research and Technology-Hellas (FORTH), Vasilika Vouton, Heraklion, GR 71110, Greece*
pnevmati@ics.forth.gr

Abstract   Intrusion Detection Systems such as Snort scan incoming packets for evidence of security threats. The computation-intensive part of these systems is a text search of packet data against hundreds of patterns, and must be performed at wire-speed. FPGAs are particularly well suited for this task and several such systems have been proposed. In this paper we expand on previous work, in order to achieve and exceed OC192 processing bandwidth (10 Gbps). We employ a scalable architecture, and use extensive fine-grained pipelining to tackle the fan-out, match, and encode bottlenecks and achieve operating frequencies in excess of 340 MHz for fast Virtex devices. To increase throughput, we use multiple comparators and allow for parallel matching of multiple search strings. We evaluate the area and latency cost of our approach and find that the match cost per search pattern character is between 4 and 5 logic cells.

Keywords: String matching, FPGA-based Network Intrusion Detection Systems, Network security

16.1   Introduction

The proliferation of Internet and networking applications, coupled with the widespread availability of system hacks and viruses have increased the need for network security. Firewalls have been used extensively to prevent access

195

P. Lysaght and W. Rosenstiel (eds.),
New Algorithms, Architectures and Applications for Reconfigurable Computing, 195–207.
© 2005 *Springer. Printed in the Netherlands.*

to systems from all but a few, well defined access points (ports), but firewalls cannot eliminate all security threats, nor can they detect attacks when they happen.

Network Intrusion Detection Systems (NIDS) attempt to detect such attempts by monitoring incoming traffic for suspicious contents. They use simple rules (or search patterns) to identify possible security threats, much like virus detection software, and report offending packets to the administrators for further actions. Snort is an open source NIDS that has been extensively used and studied in the literature [1, 2, 3, 4]. Based on a rule database, Snort monitors network traffic and detect intrusion events. An example of a Snort rule is: `alert tcp any any ->192.168.1.0/24 111(content: "idc}3a3a}"; msg: "mountd access";)`

A rule contains fields that can specify a suspicious packet's protocol, IP address, Port, content and others. The "content" (idc|3a3a|) field contains the pattern that is to be matched, written in ascii, hex or mixed format, where hex parts are between vertical bar symbols "|". Patterns of Snort V1.9.x distribution contain between one and 107 characters with an average length of 12.6 characters.

NIDS rules may refer to the header as well as to the payload of a packet. Header rules check for equality (or range) in numerical fields and are straightforward to implement. More computation-intensive is the text search of the packet payload against hundreds of patterns that must be performed at wire-speed [3, 4]. FPGAs are very well suited for this task and many such systems have been proposed [5, 6, 7]. In this paper we expand on previous work, in order to achieve and exceed a processing bandwidth of 10 Gbps, focusing on the string-matching module.

Most proposed FPGA-based NIDS systems use finite automata (either deterministic or non-deterministic) [8, 9, 6] to perform the text search. These approaches are employed mainly for their low cost, which is reported to be is between 1 and 1.5 logic elements per search pattern character. However, this cost increases when other system components are included. Also, the operation of finite automata is limited to one character per cycle operation. To achieve higher bandwidth researchers have proposed the use of packet-level parallelism [6], whereby multiple copies of the automata work on different packets at lower rate. This approach however may not work very well for IP networks due to variability of the packet size and the additional storage to hold these (possibly large) packets. Instead, in this work we employ full-width comparators for the search. Since all comparators work on the same input (one packet), it is straightforward to increase the processing bandwidth of the system by adding more resources (comparators) that operate in parallel. We evaluate the implementation cost of this approach and suggest ways to remedy the higher cost as compared to (N)DFA. We use extensive pipelining to achieve higher operating frequencies, and we address directly the fan-out of the packet to the multiple search engines, one of limiting factors identified in related work [8]. We employ a pipelined fan-out tree and achieve operating frequencies exceeding 245 MHz for VirtexE and 340 MHz for Virtex2 devices (post place & route results).

In the following section we present the architecture of our FPGA implementation, and in section 16.3 we evaluate the performance and cost of our proposed architecture. In section 16.4 we give an overview of FPGA-based string matching and compare our architecture against other proposed designs, and in section 16.5 we summarize our findings and present our conclusions.

16.2   Architecture of Pattern Matching Subsystem

The architecture of an FPGA-based NIDS system includes blocks that match the rules of the header fields and blocks that perform text matching against the entire packet payload. Of the two, the computationally expensive module is the text match. In this work we assume that it it relatively straightforward to implement the first module(s) at high speed since they involve a comparison of a few numerical fields only, and focus on making the pattern match module as fast as possible.

If the text match operates at one (input) character per cycle, the total throughput is limited by the operating frequency. To alleviate this bottleneck, other researchers suggested using packet parallelism where multiple copies of the match module scan concurrently different packets [6]. However, due to the variable size of the IP packets, this approach may not offer the guaranteed processing bandwidth. Instead, we use discrete comparators to implement a CAM-like functionality. Since each of these comparators is independent, we can use multiple instances to search for a pattern in a wider datapath. A similar approach has been used in [7].

The results of the system are (i) an indication that there was indeed a match, and (ii) the number of the rule that did match. Our architecture uses fine-grained pipelining for all sub-modules: fan-out of packet data to comparators, the comparators themselves, and for the encoder of the matching rule. Furthermore to achieve higher processing throughput, we utilize N parallel comparators per search rule, so as to process N packet bytes at the same time. In the rest of this section we expand on our design in each of these sub-modules. The overall architecture we assume is depicted in Figure 16.1. In the rest of the paper we concentrate on the text match portion of the architecture, and omit the shaded part that performs the header numerical field matching. We believe that previous work in the literature has fully covered the efficient implementation of such functions [6, 7]. Next we describe the details of the three main sub-systems: the comparators, the encoder and the fan-out tree.

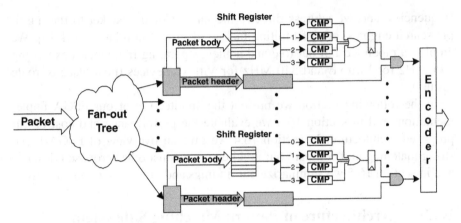

Figure 16.1. Envisioned FPGA NIDS system: Packets arrive and are fanned-out to the matching engines. N parallel comparators process N characters per cycle (four in this case), and the matching results are encoded to determine the action for this packet. The header matching logic that involves numerical field matching is shaded.

16.2.1 Pipelined Comparator

Our pipelined comparator is based on the observation that the minimum amount of logic in each pipeline stage can fit in a 4-input LUT and its corresponding register. This decision was made based on the structure of Xilinx CLBs, but the structure of recent Altera devices is very similar so our design should be applicable to Altera devices as well. In the resulting pipeline, the clock period is the sum of wire delay (routing) plus the delay of a single logic cell (one 4-input LUT + 1 flip-flop). The area overhead cost of this pipeline is zero since each logic cell used for combinational logic also includes a flip-flop. The only drawback of this deep pipeline is a longer total delay (in clock cycles) of the result. However, since the correct operation of NIDS systems does not depend heavily on the actual latency of the results, this is not a critical restriction for our system architecture. In section 16.3 we evaluate the latency of our pipelines to show that indeed they are within reasonable limits.

Figure 16.2(a) shows a pipelined comparator that matches the pattern "ABC". In the first stage comparator matches the 6 half bytes of the incoming packet data, using six 4-input LUTs. In the following two stages the partial matches are AND-ed to produce the overall match signal. Figure 16.2(b) depicts the connection of four comparators that match the same pattern shifted by zero, one, two and three characters (indicated by the numerical suffix in the comparator label). Comparator comparator_ABC+0 checks bytes 0 to 2, comparator_ABC+1 checks bytes 1 to 3 and so on. Notice that the choice of four comparators is only indicative; in general we can use N comparators, allowing the processing of N bytes per cycle.

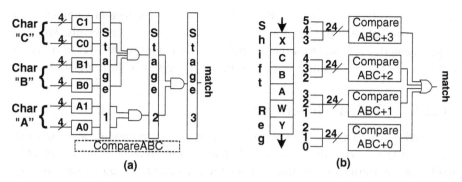

Figure 16.2. (a) Pipelined comparator, which matches pattern "ABC". (b) Pipelined comparator, which matches pattern "ABC" starting at four different offsets (+0 . . . +3).

16.2.2   Pipelined Encoder

After the individual matches have been determined, the matching rule has to be encoded and reported to the rest of the system (most likely software). We use a hierarchical pipelined encoder. In every stage, the combinational logic is described by at most 4-input, 1-output logic functions, which is permitted in our architecture.

The described encoder assumes that at most one match will occur in order to operate correctly (i.e. it is not a priority encoder). While in general multiple matches can occur in a single cycle, in practice we can determine by examining the search strings whether this situation can occur in practice. If all the search patterns have distinct suffixes, then we are ensured that we will not have multiple matches in a single cycle. However, this guarantee becomes more difficult as we increase the number of concurrent comparators. To this end we are currently working on a pipelined version of a priority encoder, which will be able to correctly handle any search string combination.

16.2.3   Packet Data Fan-out

The fan-out delay is major slow-down factor that designers must take into account. While it involves no logic, signals must traverse long distances and potentially suffer significant latencies. To address this bottleneck we created a register tree to "feed" the comparators with the incoming data. The leaves of this tree are the shift registers that feed the comparators, while the intermediate nodes of the tree serve as buffers and pipeline registers at the same time. To determine the best fan-out factor for the tree, we experimented with the Xilinx tools, and we determined that for best results, the optimal fan-out factor changes from level to level. In our design we used small fan-out for the first tree levels and increased the fan-out in the later levels of the tree up to 15 in the last tree level. Intuitively, that is because the first levels of the tree feed large blocks and the distance between the fed nodes is much larger than in last levels. We also experimented and found that the optimal fan-out from the shift-registers is 16 (15 wires to feed comparators and 1 to the next register of shift register).

16.2.4   VHDL Generator

Deriving a VHDL representation of a string matching module starting from a Snort rule is very tedious; to handle tens or hundreds of rules is not only tedious but extremely error prone. Since the architecture of our system is very regular, we developed a C program that automatically generates the desired VHDL representation directly from Snort pattern matching expressions, and we used a simple PERL script to extract all the patterns from a Snort rule file.

16.3   Evaluation Results

The quality of an FPGA-based intrusion detection system can be measured mainly using performance and area metrics. We measure performance in terms of operating frequency (to indicate the efficiency of our fine-grained pipelining) and total throughput that can be serviced, and area in terms of total area needed, as well as area cost per search pattern character.

To evaluate our proposed architecture we used four sets of rules. The first two are artificial sets that cannot be optimized (i.e. at every position all search characters are distinct), and contain 10 rules matching 10 characters each (Synth10), and 16 rules of 16 characters each (Synth16). We also used the "web-attacks.rules" from the Snort distribution, a set of 47 rules to show performance and cost for a medium size rule set, and we used the entire set of web rules (a total of 210 rules) to test the scalability of our approach for larger rule sets. The average search pattern length for these sets was 10.4 and 11.7 characters for the Web-attack and all the Web rules respectively.

We synthesized each of these rule sets using the Xilinx tools (ISE 4.2i) for several devices (the $-N$ suffix indicates speed grade): Virtex 1000–6, VirtexE 1000–8, Virtex2 1000–5, VirtexE 2600-8 and Virtex2 6000–5. The structure of these devices is similar and the area cost of our design is expected (and turns out) to be almost identical for all devices, with the main difference in performance.

16.3.1   Performance

Table 16.1 summarizes our performance results. It lists the number of rules, and the average size of the search patterns for our rule sets. It also lists the frequency we achieved using the Xilinx tools (ISE 4.2i), the percentage of wiring delay in the total delay of the critical path and the throughput (in Gbps)

Table 16.1. Performance Results: operating frequency, processing throughput, and percentage of wiring delay in the critical path.

		Synth10	Synth16	Web attacks	Web-all
	Rule Set				
	# Patterns (rules)	10	16	47	210
	Average Pattern Size (characters)	10	16	10.4	11.7
Virtex 1000 −6	MHz	193	193	171	
	Wire delay	56.7%	45.2%	61.9%	
	Gbps	6.176	6.176	5.472	
VirtexE 1000 −8	MHz	272	254	245	
	Wire delay	54.6%	57.5%	49.8%	
	Gbps	8.707	8.144	7.840	
Virtex2 1000 −5	MHz	396	383	344	
	Wire delay	37.4%	54.1%	58.7%	
	Gbps	12.672	12.256	11.008	
VirtexE 2600 −8	MHz				204
	Wire delay				70.2%
	Gbps				6.528
Virtex2 6000 −5	MHz				252
	Wire delay				69.7%
	Gbps				8.064

achieved for the design with four parallel comparators. For brevity we only list results for four parallel comparators, i.e. for processing 32 bits of data per cycle. The reported operating frequency gives a lower bound on the performance using a single (or fewer) comparators.

For our smaller synthetic rule set (labelled Synth10) we are able to achieve throughput in excess of 6 Gbps for the simplest devices and over 12 Gbps for the advanced devices. For the actual Web attack rule set (labelled 47 × 10.4), we are able to sustain over 5 Gbps for the simplest Virtex 1000 device (at 171 MHz), and about 11 Gbps for a Virtex2 device (at 344 MHz). The performance with a VirtexE device is almost 8 Gbps at 245 MHz. Since the architecture allows a single logic cell at each pipeline stage, and the percentage of the wire delay in the critical path is around 50%, it is unlikely that these results can be improved significantly.

However the results for larger rule sets are more conservative. The complete set of web rules (labelled 210 × 11.7) operates at 204 MHz and achieves a throughput of 6.5 Gbps on a VirtexE, and at 252 MHz having 8 Gbps throughput on a Virtex2 device. Since the entire design is larger, the wiring latency contribution to the critical path has increased to 70% of the cycle time. The total throughput is still substantial, and can be improved by using more parallel comparators, or possibly by splitting the design into sub-modules that can be placed and routed in smaller area, minimizing the wire distances and hence latency.

16.3.2 Cost: Area and Latency

Table 16.2 lists the total area and the area required per search pattern character (in logic cells) of rules, the corresponding device utilization, as well as the dimensions of the rule set (number of rules and average size of the search patterns). In terms of implementation cost of our proposed architecture, we see that each of the search pattern characters costs between 15 and 20 logic cells depending on the rule set. However, this cost includes four parallel comparators, so the actual cost of each search pattern character is roughly 4–5 logic cells multiplied by N for N times larger throughput.

We compute the latency of our design taking into account the three components of our pipeline: fan-out, match and encode. Since the branching factor is not fixed in the fan-out tree, we cannot offer a closed form for the number of stages. Table 16.3 summarizes the pipeline depths for the designs we have implemented: $3 + 5 + 4 = 12$ for the Synth10 and Synth16 rule set, $3 + 6 + 5 = 14$ for the Web Attacks rule set, and $5 + 6 + 7 = 18$ for the Web-all rule set. For 1,000 patterns and pattern lengths of 128 characters, we estimate the total delay of the system to be between 20 and 25 clock cycles.

We also evaluated resource sharing to reduce the implementation cost. We sorted the 47 web attack rules, and we allowed two adjacent patterns to share comparator i if their ith characters were the same, and found that the number of

Table 16.2. Area Cost Analysis.

	Synth10	Synth16	Web attacks	Web-all
# Patterns	10	16	47	210
Average Pat. Size	10	16	10.4	11.7
Virtex 1000–6	1,728 LC[1] 7%	3,896 LC 15%	8,132 LC 33%	
VirtexE 1000–8	1,728 LC 7%	3,896 LC 15%	7,982 LC 33%	
Virtex2 1000–5	1,728 LC 16%	3,896 LC 38%	8,132 LC 80%	
VirtexE 2600–8				47,686 LC 95%
Virtex2 6000–5				47,686 LC 71%
Average (LC per character)	17.28	15.21	16.9	19.40

[1]LC stands for Logic Cell, i.e. of a Slice.

Table 16.3. Pipeline Depth.

	Rule set	Synth10/ Synth16	Web attacks	Web-all	Future
	# Patterns (rules)	10/16	47	210	1,000
	Average Pattern Size (char)	10/16	10.4	11.7	
	Max Pattern Size (char)	10/16	40	62	128
Pipeline	Fan-out	3	3	5	5–10
Depth	Comparators	5	6	6	6
# Clock	Encoder	4	5	7	9
Cycles	Total	12	14	18	20–25

logic cells required to implement the system was reduced by about 30%. This is a very promising approach that reduces the implementation cost and allows more rules to be packed in a given device.

16.4 Comparison with Previous Work

In this section we attempt a fair comparison with previously reported research. While we have done our best to report these results with the most objective way, we caution the reader that this task is difficult since each system has its own assumptions and parameters, occasionally in ways that are hard to quantify.

One of the first attempts in string matching using FPGAs, presented in 1993 by Pryor, Thistle and Shirazi [11]. Their algorithm, implemented on Splash 2, succeeded in performing a dictionary search, without case sensitivity patterns, that consisted of English alphabet characters (26 characters). Pryor et al managed to achieve great performance with a low overhead AND-reduction of the match indicators using hashing.

Sidhu and Prassanna [9] used Regular Expressions and Nondeterministic Finite Automata (NFAs) for finding matches to a given regular expression. They focused on minimizing the space -$O(n^2)$- required to perform the matching, and their automata matched 1 text character per clock cycle. For a single regular expression, the NFAs and the associated FPGA circuitry were able to process each text character in 17.42-10.70ns (57.5–93.5 MHz) using a Virtex XCV100 FPGA.

Franklin, Carver and Hutchings [8] also used regular expressions to describe patterns. The operating frequency of the synthesized modules was about 30–100 MHz on a Virtex XCV1000 and 50–127 MHz on a Virtex XCV2000E, and in the order of 63.5 MHz and 86 MHz respectively on XCV1000 and XCV2000E for a few tens of rules.

Moscola, Lockwood, Loui, and Pachos used the Field Programmable Port Extender (FPX) platform, to perform string matching. They used regular expressions (DFAs) and were able to achieve operation of 37 MHz on a Virtex XCV2000E [6]. Lockwood also implemented a sample application on FPX using a single regular expression and were able to achieve an operating speed of 119 MHz on a Virtex V1000E-7 device [10].

Gokhale, et al [5] used CAMs to implement Snort rules NIDS on a Virtex XCV1000E. Their hardware runs at 68MHz with 32-bit data every clock cycle, giving a throughput of 2.2 Gbps, and reported a 25-fold improvement on the speed of Snort v1.8 on a 733 MHz PIII and an almost 9-fold improvement on a 1 GHz PowerPC G4.

Clark and Schimmel [12] developed a pattern matching coprocessor that supports the entire SNORT ruleset using NFAs. In order to reduce design area they used centralized decoders instead of character comparators for the NFA state transitions. They stored over 1,500 patterns (17,537 characters) in a Virtex-1000. Their design required about 1.1 logic cells per matched character, ran at 100 MHz and processed one character per clock cycle, with a throughput 0.8 Gbps.

The recent work by Cho, Navab and Mangione-Smith [7] is closer to our work. They designed a deep packet filtering firewall on a FPGA and automatically translated each pattern-matching component into structural VHDL. The micro-architecture of the content pattern match unit used 4 parallel comparators for every pattern so that the system advances 4 bytes of input packet every clock cycle. The design implemented in an Altera EP20K device runs at 90 MHz, achieving 2.88 Gbps throughput. They require about 10 logic cells per search pattern character. However, they do not include the fan-out logic that we have, and do not encode the matching rule. Instead they just OR all the match signals to indicate that some rule matched.

The results of these works are summarized in the bottom portion of Table 1, and we can see that most previous works implement a few tens of rules at most, and achieve throughput less than 4 Gbps. Our architecture on the same or equivalent devices achieves roughly twice the operating frequency and throughput. In terms of best performance, we achieve 3.3 times better processing throughput compared with the fastest published design which implements a single search pattern. Our 210-rule implementation achieves at least a 70% improvement in throughput compared to the fastest existing implementation.

16.5   Conclusions and Future Work

We have presented an architecture for Snort rule match in FPGAs. We propose the use of extensive fine-grained pipelining in order to achieve high

Table 16.4. Detailed comparison of string matching FPGA designs.

Description	Input Bits/ c.c.	Device	Freq. MHz	Throughput (Gbps)	Logic Cells[1]	Logic Cells /char	#Patterns × #Characters
Sourdis-	32	Virtex	193	6.176	1,728	17.28	10 × 10
Pnevmatikatos		1000	171	5.472	8,132	16.64	47 × 10.4
Discrete		VirtexE	272	8.707	1,728	17.28	10 × 10
Comparators		1000	245	7.840	7,982	16.33	47 × 10.4
		Virtex2	396	12.672	1,728	17.28	10 × 10
		1000	344	11.008	8,132	16.64	47 × 10.4
		VirtexE 2600	204	6.524	47,686	19.40	210 × 11.7
		Virtex2 6000	252	8.064	47,686	19.40	210 × 11.7
Sidhu et al.[9]	8	Virtex	93.5	0.748	280	~31	$(1\times)\,9^4$
NFAs		100	57.5	0.460	1,920	~66	$(1\times)\,29^4$
Franklin et al.[8]	8	Virtex	31	0.248	20,618	2.57	8,003[5]
Regular		1000	99	0.792	314	3.17	99
Expressions			63.5	0.508	1726	3.41	506
		VirtexE	50	0.400	20,618	2.57	8,003
		2000	127	1.008	314	3.17	99
			86	0.686	1726	3.41	506
Moscola et al.[6] DFAs[7]	32	VirtexE 2000	37	1.184	8,134[2]	19.4	21×20^3
Lockwood[10] FSM+counter	32	VirtexE 1000	119	3.808	98	8.9	1 × 11
Gokhale et al.[5] Dis. Comparators	32	VirtexE 1000	68	2.176	9,722	15.2	32 × 20
Cho et al.[7] Dis. Comparators	32	Altera EP20K	90	2.880	~17,000	10.55	105 × 15.3
Clark et al.[12] NFAs Shared Decoders	8	Virtex 1000	100	0.800	~19,660	~1.1	$1,500 \times 11, 7^6$

[1]Two *Logic Cells* form one *Slice* and:

- 2 Slices form one *CLB* (Virtex-VirtexE devices).
- 4 Slices form one *CLB* (in Virtex2-Virtex2 Pro Devices).

[2]These results do not include the cost/area of infrastructure and protocol wrappers.
[3] 21 regular expressions, with 20 characters on average, (about 420 character).
[4]One regular expression of the form $(a \mid b)^*a(a \mid b)^k$ for k = 8 and 28. Because of the * operator the regular expression can match more than 9 or 29 characters.
[5]Sizes refer to Non-meta characters and are roughly equivalent to 800, 10, and 50 patterns of 10 characters each.
[6]Over 1,500 patterns that contain 17,537 characters.
[7]4 Parallel FSMs on different Packets.

operating frequencies, and parallel comparators to increase the processing throughput. This combination proves very successful, and the throughput of our design exceeded 11 Gbps for about 50 Snort rules. These results offer a distinct step forward compared to previously published research. If latency is not critical to the application, fine-grained pipelining is very attractive in FPGA-based designs: every logic cell contains one LUT and one flip-flop, hence the pipeline area overhead is zero. The current collection of Snort rules contains fewer than 1500 patterns, with an average size of 12.6 characters. Using the area cost as computed earlier, we need about 3 devices of 120,000 logic cells to include the entire Snort pattern matching and about 4 devices to include the entire snort rule set including header matching. These calculations do not include area optimizations, which can lead to further significant improvements.

Throughout this paper we used four parallel comparators. However, a different level of parallelism can also be used depending on the bandwidth demands. Reducing the processing width leads to a smaller, possibly higher frequency design, while increasing the processing width leads to a bigger and probably lower frequency design. Throughput depends on both frequency and processing width, so we need to seek the cost effective tradeoff of these two factors.

Despite the significant body of research in this area, there are still improvements that we can use to seek better solutions. Our immediate goals are to use hierarchical decomposition of large rule set designs and attempt to use multiple clock domains. The idea is to use a slow clock to drive long but wide busses to distribute data and a fast clock for local processing that only uses local wiring. The target would be to retain the frequency advantage of our medium-sized design (47 rules) for a much larger rule set. This technique is feasible since all the devices we used in this paper as well as their placement and routing tools support multiple clock domains.

Furthermore, the use of decoding before the actual character matching, an approach similar to the one used by Clark and Schimmel [12], would be useful to reduce area cost. In this approach the characters are decoded before the comparators that perform the matching function. Then the comparison consists simply of selecting the proper decoded signal and matching a N character string is achieved simply with a N-input AND gate. According to our preliminary results, shared decoders can lower our design's area cost to one quarter of the original. When combined with the use of multiple clock domains for data distribution and processing it can achieve a performance similar to the one we achieved in our full comparator architecture.

Acknowledgments

This work was supported in part by the IST project SCAMPI (IST-2001-32404) funded by the European Union.

References

[1] SNORT official web site: (http://www.snort.org)
[2] Roesch, M.: Snort—lightweight intrusion detection for networks. In: Proceedings of LISA'99: 13th Administration Conference. (1999) Seattle Washington, USA.
[3] Desai, N.: Increasing performance in high speed NIDS. In: www.linuxsecurity.com. (2002)
[4] Coit, C.J., Staniford, S., McAlerney, J.: Towards faster string matching for intrusion detection or exceeding the speed of snort. In: DISCEXII, DAPRA Information Survivability conference and Exposition. (2001) Anaheim, California, USA.
[5] Gokhale, M., Dubois, D., Dubois, A., Boorman, M., Poole, S., Hogsett, V.: Granidt: Towards gigabit rate network intrusion detection technology. In: Proceedings of 12th International Conference on Field Programmable Logic and Applications. (2002) France.
[6] Moscola, J., Lockwood, J., Loui, R.P., Pachos, M.: Implementation of a content-scanning module for an internet firewall. In: Proceedings of IEEE Workshop on FPGAs for Custom Computing Machines. (2003) Napa, CA, USA.
[7] Young, H. Cho, S.N., Mangione-Smith, W.: Specialized hardware for deep network packet filtering. In: Proceedings of 12th International Conference on Field Programmable Logic and Applications. (2002) France.
[8] Franklin, R., Carver, D., Hutchings, B.: Assisting network intrusion detection with reconfigurable hardware. In: IEEE Symposium on Field-Programmable Custom Computing Machines. (2002)
[9] Sidhu, R., Prasanna, V.K.: Fast regular expression matching using fpgas. In: IEEE Symposium on Field-Programmable Custom Computing Machines. (2001) Rohnert Park, CA, USA.
[10] Lockwood, J.W.: An open platform for development of network processing modules in reconfigurable hardware. In: IEC DesignCon '01. (2001) Santa Clara, CA, USA.
[11] Pryor, D.V., Thistle, M.R., Shirazi, N.: Text searching on splash 2. In: Proceedings of IEEE Workshop on FPGAs for Custom Computing Machines. (1993) 172–177.
[12] Clark, C.R., Schimmel, D.E.: Efficient Reconfigurable Logic Circuit for Matching Complex Network Intrusion Detection Patterns. In: Proceedings of 13th International Conference on Field Programmable Logic and Applications (Short Paper). (2003), Lisbon, Portugal.

Chapter 17

Architecture and FPGA Implementation of a Digit-serial RSA Processor

Alessandro Cilardo[1], Antonino Mazzeo[2], Luigi Romano[3], Giacinto Paolo Saggese[4]

[1] *Dipartimento di Informatica e Sistemistica, Universita' degli Studi di Napoli*
acilardo@unina.it

[2] *Dipartimento di Informatica e Sistemistica, Universita' degli Studi di Napoli*
mazzeo@unina.it

[3] *Dipartimento di Informatica e Sistemistica, Universita' degli Studi di Napoli*
lrom@unina.it

[4] *Dipartimento di Informatica e Sistemistica, Universita' degli Studi di Napoli*
saggese@unina.it

Keywords: Field-Programmable Gate Array (FPGA), RSA cryptosystem, Modular Multiplication, Modular Exponentiation

Introduction

In the recent years, we have witnessed an increasing deployment of hardware devices for providing security functions via cryptographic algorithms. In fact, hardware devices provide both high performance and considerable resistance to tampering attacks, and are thus ideally suited for implementing computationally intensive cryptographic routines which operates on sensitive data.

Among the various techniques found in the cryptography realm, the Rivest-Shamir-Adleman (RSA) algorithm [1] constitutes the most widely adopted public-key scheme. In particular, it is useful for security applications which need confidentiality, authentication, integrity, and non-repudiation [2]. The basic operation of this algorithm is modular exponentiation on large integers,

P. Lysaght and W. Rosenstiel (eds.),
New Algorithms, Architectures and Applications for Reconfigurable Computing, 209–218.
© 2005 *Springer. Printed in the Netherlands.*

i.e. $Y = X^E \bmod N$, which is used for both decryption/signature and encryption/verification.

The security level of an RSA cryptosystem is tied to the length of the modulus N. Typical values of the modulus size range from 768 to 4096 bits, depending on the security level required. All operands involved in the computation of modular exponentiation have normally the same size as the modulus.

All existing techniques for computing $X^E \bmod N$ reduce modular exponentiation to a sequence of modular multiplications. Several sub-optimal algorithms have been presented in the literature to compute the sequence of multiplications leading to the Eth power of X, such as binary methods (RL-algorithm and LR-algorithm), M-ary methods, Power Tree, and more [3, 4]. In particular, the method known as Binary Right-to-Left Algorithm [4] consists of repeated squaring and multiplication operations and is well-suited for simple and efficient hardware implementations.

Since modular multiplication is the core computation of all modular exponentiation algorithms, the efficiency of its execution is crucial for any implementation of the RSA algorithm. Unfortunately, modular multiplication is a complex arithmetic operation because of the inherent multiplication and modular reduction operations. Several techniques have been proposed in the last years for achieving efficient implementations of modular multiplication. In particular, Blakley's method [5] and Montgomery's method [6] are the most studied techniques. Actually, they are the only algorithms suitable for practical hardware implementation [3].

Both Blakley's method and Montgomery's method perform the modular reduction during the multiplication process. No division operation is needed at any point in the process. However, Blakley's method needs a comparison of two large integers at each step of the modular multiplication process, while the Montgomery's method does not, by means of a representation of the operands as a residue class modulo N. Furthermore, the Montgomery's technique requires some preprocessing and postprocessing steps, which are needed to convert the numbers to and from the residue based representation. However, the cost of these steps is negligible when many consecutive modular multiplications are to be executed, as in the case of RSA. This is the reason why the Montgomery's method is considered the most efficient algorithm for implementing RSA operations. There exist several versions of the Montgomery's algorithm, depending on the number r used as the radix for the representation of numbers. In hardware implementations r is always a power of 2.

The Montgomery's algorithm is in turn based on repeated additions on integer numbers. Since the size of operands is as large as the modulus, the addition operation turns out to be the critical step from the implementation viewpoint.

In this paper we present an innovative hardware architecture and an FPGA-based implementation of the Montgomery's algorithm based on a digit-serial approach, which allows the basic arithmetic operations to be broken into words and processed in a serialized fashion. As a consequence, the architecture implementation takes advantage of a short critical path and low area requirements. In fact, as compared to other solutions in the literature, the proposed implementation of the RSA processor has smaller area requirements and comparable performance. The serialization factor S of our serial architecture is taken as a parameter and the final performance level is given as a function of this factor. We thoroughly explore the design tradeoffs, in terms of area requirements vs time performance, for different values of the key length and the serialization factor.

17.1 Algorithm Used for the RSA Processor

This section describes the algorithms we have used in our RSA processor. For implementation of modular multiplication we exploit some optimizations of the Montgomery Product first described by Walter [7]. We assume that N can be represented with K bits, and we take $R = 2^{K+2}$. The N-residue of A with respect to R is defined as the positive integer $\bar{A} = A \cdot R \bmod N$. The Montgomery Product [6] of residues of A and B, *MonPro*$(\bar{A}, \bar{B})$, is defined as $(\bar{A} \cdot \bar{B} \cdot R^{-1}) \bmod N$, that is the N-residue of the desired $A \cdot B \bmod N$. If $A, B < 2N$, combining [7] and [3], the following radix-2 binary add-shift algorithm can be employed to calculate *MonPro*:

Algorithm 17.1.1— *Montgomery Product MonPro(A,B) radix-2.*
Given $A = \sum_{i=0}^{K+2} A_i \cdot 2^i$, $B = \sum_{i=0}^{K} B_i \cdot 2^i$, $N = \sum_{i=0}^{K-1} N_i \cdot 2^i$, where $A_i, B_i, N_i \in \{0, 1\}$, $A_{K+1}, A_{K+2} = 0$, computes a number falling in $[0, 2N]$ which is modulo N congruent with desired $(A \cdot B \cdot 2^{-(K+2)}) \bmod N$
1. $U = 0$
2. For $j = 0$ to $K + 2$ do
3.   if $(U_0 = 1)$ then $U = U + N$
4. $U = (U/2) + A_j \cdot B$
5. end for

We report the exponentiation algorithm for computing $X^E \bmod N$ known as Right-To-Left binary method [4], modified here in order to take advantage of the Montgomery Product.

Algorithm 17.1.2— *Right-To-Left Modular Exponentiation using Montgomery Product.*
Given X, N, and $E = \sum_{i=0}^{H-1} E_i \cdot 2^i$, $E_i \in \{0, 1\}$, computes $P = X^E \bmod N$.

1. $P_0 = MonPro(1, R^2 \bmod N)$
2. $Z_0 = MonPro(X, R^2 \bmod N)$
3. For $i = 0$ to $H - 1$ do
4. $Z_{i+1} = MonPro(Z_i, Z_i)$
5.   if $(E_i = 1)$ then $P_{i+1} = MonPro(P_i, Z_i)$
6.   else $P_{i+1} = P_i$
7. end for
8. $P = MonPro(P_H, 1)$
9. if $(P \geq N)$ then return $P - N$
10.   else return P

The first phase (lines 1–2) calculates the residues of the initial values 1 and X. For a given key value, the factor $R^2 \bmod N$ remains unchanged. It is thus possible to use a precomputed value for such a factor and reduce residue calculation to a *MonPro*. The core of the computation is a loop in which modular squares are performed, and the previous partial result P_i is multiplied by Z_i, based on a test performed on the value of i-th bit of E (H is the number of the bits composing E). It is worth noting that, because of the absence of dependencies between instructions 4 and 5–6, these can be executed concurrently. Instruction 8 allows to switch back from the residue domain to the normal representation of numbers. After line 8 is executed, a further check is needed (lines 9–10) to ensure that the obtained value of P is actually $X^E \bmod N$. In fact, while it is acceptable in intermediate loop executions that the *MonPro* temporary result (line 4 and line 5) be in the range $[0, 2N]$, this cannot be in the last iteration. Thus, if the final value of P is greater than N, it must be diminished by N.

17.2   Architecture of the RSA Processor

As shown in Algorithm 17.1.2, the modular multiplication constitutes the basic operation of the RSA exponentiation, which basically consists of repeated multiplication operations. Thus, the RSA processor must be able to properly sequence modular products on data and store intermediate results in a register file according to the control flow of Algorithm 17.1.2. In turn, the *MonPro* operation (see Algorithm 17.1.1) is composed of different micro-operations consisting of load/store on registers, shifts and additions. All operands are $(K + 3)$-bit long at most, where K is the modulus length. In our implementation, each arithmetic operation is broken up into a number of steps and executed in a serialized fashion on S-bit words.

At a high level of abstraction, the RSA processor is composed of two modules (see Figure 17.1): a *Data-path Unit* performing data-processing operations, and a *Control Unit* which determines the sequence of operations.

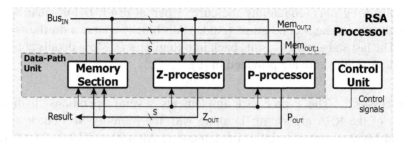

Figure 17.1.   Overall architecture of RSA processor.

The *Control Unit* reflects the structure of Algorithm 17.1.2, in that it is composed of two hierarchical blocks corresponding to the main modular exponentiation routine of Algorithm 17.1.2 and the *MonPro* subroutine of Algorithm 17.1.1, respectively.

More precisely, a block named *MonExp_Controller* is in charge of generating the control flow signals (for loops, conditional and unconditional jumps), activating a second block (the *MonPro_Controller* block) when a modular product is met, and waiting until this has finished. The *MonPro_Controller* supervises modular product execution, i.e. it sequences long integer operations, which are performed serially on S-bit words.

The *Data-path* has a data width of S bits. Thus, S represents the serialization factor or, in other terms, S is the digit-size in multiprecision arithmetic. Data-path is composed of three macro blocks (see Figure 17.2): a *Memory Section* block storing intermediate data inherent in Algorithm 17.1.2, and two processing units named *P-processor* and *Z-processor*, executing concurrently modular products and modular squaring operations, respectively.

The data-path is organized as a 3-stage pipeline. The first stage fetches the operands. Data are read from the register file in the Memory Section and from

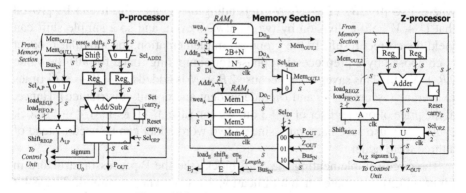

Figure 17.2.   Structure of the Data-path.

scratchpad registers, containing the current partial result U for squaring and multiplication. The second stage (Adder/Sub block) operates on the fetched data. The last stage writes results back into Registers U. More details about the presented architecture are provided in [9].

P-processor. The P-processor implements several operations in distinct phases of the RSA algorithm: 1) along with the Memory Section, it acts as a serial Montgomery multiplier implementing Algorithm 17.1.1; 2) it carries out a simple preprocessing phase to accelerate modular multiplication (i.e. $2B$ and $2B + N$ computation); 3) finally, it performs the reduction step (a comparison and a conditional subtraction) to ensure that the result is actually the modulus of the requested exponentiation.

MonPro(A, B) of Algorithm 17.1.1 is a sequence of $K + 3$ conditional sums, in which the operands depend on both the least significant bit U_0 of the partial result U, and the ith bit of the operand A. The $(K + 2)$-bit additions are performed serially with an S-bit adder in $M = \lceil (K + 2)/S \rceil$ clock cycles. This has two fundamental advantages. First, it allows area saving, provided that the area-overhead due to serial to parallel data formatting and the subsequent inverse conversion does not frustrate area-saving deriving from smaller operands and a narrower data-path. Second, a serial approach avoids long carry chains, which are difficult to implement in technologies such as FPGA, and result in larger net delays. These advantages come at the price of the time-area overhead due to serial-to-parallel (and vice versa) conversions. By addressing the RAM as a set of register files and accessing directly the required portion of operands, we can get rid of multiplexer/demultiplexer blocks. This results in a dramatic reduction of time-area overhead.

The computational core of the P-processor corresponds to Steps 3–4 of Algorithm 17.1.1 and is actually implemented as:

3 . $U = (U + A_i \cdot 2B + U_0 \cdot N)/2 = (U + V)/2$

where $V \in \{0, 2B, 2B + N, N\}$ depending on A_i and U_0. It can be proved that $U + V$ before division by two is always even, and so a simple shift can yield the correct result. In the proposed implementation, the modular product is composed by a preprocessing phase for computing (once for all) the value $2B + N$ [8]. This saves time, because $2B + N$ is added to U $(K + 3)/4$ times in average (assuming A_i and U_0 independent and equally distributed). Hence, at the price of one register of $K + 3$ bit, we can save $M \cdot ((K + 3)/4 - 1)$ clock cycles, and also save hardware, since a two word adder can be used instead of a three word adder.

The circuit implementing the *Shift* block (see the P-processor structure in Figure 17.2), computes a multiplication by two when required. The S-bit registers *Reg* are pipelining registers. The *Adder/Subtracter* sums operands, or subtracts the operand on the right input from the one on the left input. Note that the Adder/Subtracter performs multiprecision operations. *Register U* stores the value of U prior to shifting, as required by the modular product algorithm. It shows its contents shifted down by 1 bit. It also outputs the least significant bit of U, necessary for the *Controller* to choose which is the next operand that is to be added to U. *Register A* holds the $(K + 3)$-bit value of the A operand of modular product, which can be loaded serially from the Memory Section. It can shift down one bit at a time, showing the A_i bit to the controller.

Memory Section.   The Memory Section schematics is reported in Figure 17.2. Memory Section supplies P and Z processors with the contents of the register file, in a serial fashion, S bits at a time. It receives output data from processors or from external through Bus_{IN}. Memory RAM_0 stores $(K + 3)$-bit numbers that can be added to the partial sum U of Algorithm 17.1.1 (P, Z, $2B + N$, N), while RAM_1 stores K-bit constants that are useful for calculating an N-residue, or returning to normal number representation. Each operand is stored as an array of S-bit words, in order to avoid the use of area-consuming multiplexers for selecting the correct part of the operand to be summed. The Memory Section also contains *Register E*, which stores the exponent E.

Z-Processor.   The Z-processor (see Figure 17.2) is a simplified version of the P-Processor, intended to perform a *MonPro* operation concurrently with the P-Processor. Note that, strictly speaking, we only need a P-processor and a Memory-Section to implement modular product and squaring. Hence, the Z-processor could be discarded (at the cost of doubling the time for modular exponentiation). This simplification however scales down the overall required area by a factor smaller than 2 (area of Memory Section is constant), so the version of the RSA processor with both the P and the Z processors is characterized by a better value of the product $A \cdot T$. When the available area is reduced and performance is not critical, the design option which relies solely on the P-processor can gain interest.

17.3   FPGA Implementation and Performance Analysis

We have used a Xilinx Virtex-E 2000-8bg560 for implementing the proposed architecture. Xilinx XCV2000E has 19200 slices and 19520 tristate buffers. For synthesis we used Synplicity Synplify Pro 7.1, integrated in Xilinx ISE 4.1.

We implemented our design in RTL VHDL for $K = 1024$ and evaluated it for values of the widths S ranging from 32 to 256. First we verified the correctness of the design. We then performed synthesis, place&route step, and

S	Stage	T_{ck} [ns]	Total Slices	FF/LUT slices	Lut for Dual-port RAM/Single-port RAM	Lut for Shift Reg	Tristate buffers	Mux implemen-tation
32	Memory	8,7	627	44 / 486	512 / 256	0	0	LUTs
	Adder	7,4	32	2 / 64	0 / 0	0	0	
	Write	8,6	336	385 / 143	256 / 0	128	448	Tristates
64	Memory	10,7	563	44 / 108	512 / 256	0	960	Tristates
	Adder	11,6	64	2 / 128	0 / 0	0	0	
	Write	11,5	561	577 / 209	512 / 0	128	1152	Tristates
128	Memory	17,5	688	44 / 34	512 / 256	0	1344	Tristates
	Adder	18,6	132	2 / 264	0 / 0	0	0	
	Write	17,8	1050	1025 / 278	1024 / 0	256	2816	Tristates
256	Memory	30,5	553	44 / 30	512 / 256	0	2122	Tristates
	Adder	30,9	264	2 /528	0 / 0	0	0	
	Write	30,2	2085	2049 / 543	2048 / 0	512	4608	Tristates

Figure 17.3. Hardware resources and clock periods for each pipeline stage varying S.

timing verification. A partial manually floorplan of individual blocks was carried out occasionally, upon need.

Figure 17.3 reports the minimum clock period and the total number of slices required for different values of S and for each stage of the pipeline. Results show that the stage which limits the clock frequency is the Adder. Hence, for a fixed S, we determined the maximum sustainable clock rate from the Adder, and used that as the target clock for other stages. We exploited the capability of Virtex devices to implement multiplexers using tristate buffers instead of LUTs. From a detailed analysis of the architecture, it follows that the number of clock cycles for each modular product is given by $(2M + 2) + (K + 3) \cdot M$. The number of sequential modular products is $H + 2$ (where H is the number of bits composing the exponent E) because squarings and products of the modular exponentiation loop are executed in parallel, and a product is necessary for residue calculation and for turning the residue of the result in normal representation. Finally, M clock cycles are needed for subtraction, and $M + 1$ clock cycles are needed for the last S bits of the result to appear on the output P_{out}. The number of clock cycles (N_{CK}), the total area, and the total time for the exponent $E = 2^{16} + 1$ are reported in Figure 17.4 as functions of parameter S. These results are contrasted against other implementations in the following section.

S	T_{ck} [ns]	Area [Slices]	N_{CK}	Total Time [ms]	A·T [Slices·ms]
32	8,74	995	645288	5,64	5612
64	11,6	1188	176016	3,86	4586
128	18,6	1870	332440	3,27	6115
256	30,9	2902	97804	2,99	8677

Figure 17.4. RSA encryption with 1024 bit key.

17.4   Related Work

Most hardware implementations of modular exponentiation are either dated or they rely on an ASIC implementation. As a consequence, a fair comparison is difficult. In [7] the Montgomery technique and the systolic multiplication were combined for the first time, resulting in a bidimensional systolic array architecture which gave a throughput of one modular multiplication per clock cycle and a latency of $2K + 2$ cycles. The great area requirements constituted the major drawback of this structure, deriving from its inherent high-parallelism.

A unidimensional systolic architecture was proposed in [8]. This work implements the same algorithm as ours (radix-2 Montgomery algorithm for modular product Algorithm 17.1.1), on top of a Xilinx device with the same basic architecture, but with a different technology, namely a XC40250XV-09 device. In a successive study [10], Blum and Paar improved their architecture using a high-radix formulation of Algorithm 17.1.1. To output the entire result, [8] requires $2(H + 2)(K + 4) + K/U$ clock cycles, where U is the dimension of the processing elements. The fastest design ($U = 4$) of [8] requires 0.75 ms for the same encryption of Figure 17.4 and requires 4865 XC4000 CLBs that are equivalent to 4865 Virtex slices. Our fastest design ($S = 256$) requires 2.99 ms (4 times slower), but it requires 2902 slices (with a saving of area equal to 40%). Our design requiring the least area ($S = 32$) occupies only 995 slices, while the smallest one in [8] requires 3786 slices. Finally, our design with the best $A \cdot T$ product ($S = 64$) presents $A \cdot T = 4586$, while the corresponding design of [8], presents $A \cdot T = 3511$.

In summary, the solution presented in [8] exhibits better performance, as compared to ours. This was made possible by the improved parallelism due to pipelining, inherent in the systolic paradigm of computation. On the other hand, our implementation is slower, but it has lower area requirements. This would allow the RSA processor to be replicated on the same physical device and more modular exponentiation operations to be performed in parallel.

17.5   Conclusions

We presented a novel serial architecture for RSA encryption/decryption operation. The design is targeted for implementation on reconfigurable logic, and exploits the inherent characteristics of the typical FPGA devices. The design allows to tradeoff area for performance, by modifying the value of the serialization factor S. Quantitative evaluation of this tradeoff was conducted via simulation. The results obtained showed that the presented architecture achieves good performance with low area requirements, as compared to other architectures.

218

Acknowledgements

This work was partially funded by Regione Campania within the framework of the Telemedicina and Centri di Competenza Projects.

References

[1] R. L. Rivest et al., "A Method for Obtaining Digital Signatures", Commun. ACM, vol. 21, pp. 120–126, 1978.

[2] A. Menezes, P. van Oorschot, and S. Vanstone, Handbook of Applied Cryptography, CRC Press, 1996.

[3] Ç. K. Koç, "High-speed RSA Implementation", Technical Report TR 201, RSA Laboratories, November 1994.

[4] D. E. Knuth, "The Art of Computer Programming: Seminumerical Algorithms", vol. 2, Addison-Wesley, 1981.

[5] G. R. Blakley, "A computer algorithm for the product . . . ", IEEE Trans. on Computers, Vol. 32, No. 5, pp. 497–500, May 1983.

[6] P. L. Montgomery, "Modular multiplication without trial division", Math. of Computation, 44(170):519–521, April 1985.

[7] C. D. Walter, "Systolic Modular Multiplication", IEEE Trans. on Computers, Vol. 42, No. 3, pp. 376–378, March 1993.

[8] T. Blum, and C. Paar, "Montgomery Modular Exponentiation . . . ", Proc. 14th Symp. Comp. Arith., pp. 70–77, 1999.

[9] A. Mazzeo, N. Mazzocca, L. Romano, and G. P. Saggese, "FPGA-based Implementation of a Serial RSA processor", Proceedings of the Design And Test Europe (DATE) Conference 2003, pp. 582–587.

[10] T. Blum, and C. Paar, "High-Radix Montgomery Modular . . . ", IEEE Trans. on Comp., Vol. 50, No. 7, pp. 759–764, July 2001.

Chapter 18

Division in *GF(p)* for Application in Elliptic Curve Cryptosystems on Field Programmable Logic

Alan Daly[1], William Marnane[1], Tim Kerins[1], and Emanuel Popovici[2]

[1] *Department of Electrical & Electronic Engineering, University College Cork, Ireland*
Email: {aland,liam,timk}@rennes.ucc.ie

[2] *Department of Microelectronic Engineering, University College Cork, Ireland*
Email: e.popovici@ucc.ie

Abstract   Elliptic Curve Cryptosystems (ECC) are becoming increasingly popular for use in mobile devices and applications where bandwidth and chip area are limited. They provide much higher levels of security per key bit than established public key systems such as RSA. The core ECC operation of point scalar multiplication in *GF(p)* requires modular multiplication, division/inversion and addition/subtraction. Division is the most costly operation in terms of speed and is often avoided by performing many extra multiplications. This paper proposes a new divider architecture and FPGA implementations for use in an ECC processor.

18.1   Introduction

Elliptic Curve Cryptosystems (ECC) were independently proposed in the mid-eighties by Victor Miller [Miller, 1985] and Neil Koblitz [Koblitz, 1987] as an alternative to existing public key systems such as RSA and DSA. No sub-exponential algorithm is known to solve the discrete logarithm problem on a suitably chosen elliptic curve, meaning that smaller parameters can be used in ECC with equivalent security to other public key systems. It is estimated that an elliptic curve group with 160-bit key length has security equivalent to RSA with a key length of 1024-bit [Blake et al., 2000].

Two types of finite field are popular for use in elliptic curve public key cryptography: *GF(p)* with *p* prime, and $GF(2^n)$ with *n* a positive integer. Many implementations focus on using the field $GF(2^n)$ due to the underlying arithmetic

219

P. Lysaght and W. Rosenstiel (eds.),
New Algorithms, Architectures and Applications for Reconfigurable Computing, 219–229.

which is well suited to binary numbers [Ernst et al., 2002]. However, most $GF(2^n)$ processors are limited to operation on specified curves and key sizes. An FPGA implementation of a $GF(2^n)$ processor which can operate on different curves and key sizes without reconfiguration has previously been presented in [Kerins et al., 2002]. ECC standards define different elliptic curves and key sizes which ECC implementations must be capable of utilising [IEEE, 2000]. With a $GF(p)$ processor, any curve or key length up to the maximum size p, can be used without reconfiguration.

Few implementations of ECC processors over $GF(p)$ have been implemented in hardware to date due to the more complicated arithmetic required [Orlando and Paar, 2001].

The modular division operation has been implemented in the past by modular inversion followed by modular multiplication. No implementations to date have implemented a dedicated modular division component in an ECC application.

This paper proposes a modular divider for use in an ECC processor targeted to FPGA which avoids carry chain overflow routing. The results presented include divider bitlengths of up to 256-bits, which would provide security well in excess of RSA-1024 when used in an ECC processor.

18.2   Elliptic Curve Cryptography over $GF(p)$

An elliptic curve over the finite field $GF(p)$ is defined as the set of points (x, y), which satisfy the elliptic curve equation

$$y^2 = x^3 + ax + b$$

where x, y, a and b are elements of the field and $4a^3 + 27b^2 \neq 0$.

To encrypt data, it is represented as a point on the chosen curve over the finite field. The fundamental encryption operation is point scalar multiplication, i.e. a point P (x_P, y_P) is added to itself k times, to get point Q (x_Q, y_Q).

$$\begin{aligned} Q &= kP \\ &= \underbrace{P + P + \cdots + P}_{k \text{ times}} \end{aligned}$$

Recovery of k, through knowledge of the Elliptic Curve equation, base point P, and end point Q, is called the Elliptic Curve Discrete Logarithm Problem (ECDLP) and takes fully exponential time to achieve. The Elliptic Curve Diffie-Hellman Key Exchange protocol uses this fact to generate shared secret keys used for bulk data encryption.

In order to compute kP, a double and add method is used and k is represented in binary form and scanned right to left from LSB to MSB, performing a double

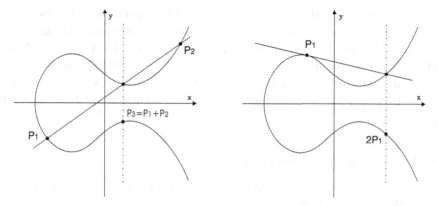

Figure 18.1.   Point Addition.   *Figure 18.2.*   Point Doubling.

at each step and an addition if k_i is 1. Therefore the multiplier will require $(m - 1)$ point doublings and an average of $(\frac{m-1}{2})$ point additions, where m is the bitlength of the field prime p.

The operations of elliptic curve point addition and point doubling are best explained graphically as shown in Fig. 18.1 and Fig. 18.2. To add two distinct points, P_1 and P_2, a chord is drawn between them. This chord will intersect the curve at exactly one other point, and the reflection of that point through the x-axis is defined to be the point $P_3 = P_1 + P_2$.

In the case of adding point P_1 to itself (doubling P_1), the tangent to the curve at P_1 is drawn and found to intersect the curve again at exactly one other point. The reflection of this point through the x-axis is defined to be the point $2P_1$. (Note: The *point at infinity*, $\mathcal{O}$ is taken to exist infinitely far on the y-axis, and is the identity of the elliptic curve group.)

The point addition/doubling formulae in *affine coordinates* are given below. Let $P_1 = (x_1, y_1)$ and $P_2 = (x_2, y_2)$, then $P_3 = (x_3, y_3) = P_1 + P_2$ is given by:

$$x_3 = \lambda^2 - x_1 - x_2$$
$$y_3 = \lambda(x_1 - x_3) - y_1$$
$$\lambda = \begin{cases} \dfrac{y_2 - y_1}{x_2 - x_1} & \text{if } P_1 \neq P_2 \\[2mm] \dfrac{3x_1^2 + a}{2y_1} & \text{if } P_1 = P_2 \end{cases}$$

Different coordinate systems can be used to represent the points on an elliptic curve, and it is possible to reduce the number of inversions by representing the points in *projective coordinates*. However, this results in a large increase in the number of multiplications required per point operation (16 for addition, 10 for doubling) [Blake et al., 2000].

Processors using affine coordinates require far fewer clock cycles and hence less power than processors based on projective coordinates [Orlando and Paar, 2001]. Point addition requires 2 multiplications, 1 division and 6 addition/subtraction operations. Point doubling requires 3 multiplications, 1 division and 7 additions/subtractions. The overall speed of an affine coordinate-based ECC processor will hence rely on an efficient modular division operation.

18.3 Modular Inversion

A common method to perform the division operation is to perform an inversion followed by a multiplication.

The multiplicative inverse of an integer a (mod M) exists if and only if a and M are relatively prime. There are two common methods to calculate this inverse. One is based on Fermat's theorem which states that $a^{M-1}(\text{mod } M) = 1$ and therefore $a^{M-2}(\text{mod } M) = a^{-1} (\text{mod } M)$. Using this fact, the inverse may be calculated by modular exponentiation. However this is an expensive operation requiring on average $1.5 \log_2 m$ multiplications.

Kaliski proposed a variation of the extended Euclidean algorithm to compute the Montgomery inverse, $a^{-1}2^m$ (mod M) of an integer a [Kaliski, 1995]. The output of this algorithm is in the Montgomery domain, which is useful when performing further Montgomery multiplications. The Montgomery multiplication algorithm is an efficient method to perform modular multiplication without the need for trial division. When many multiplications are necessary, its benefits are fully exploited by mapping inputs into the "Montgomery Domain" before multiplication, and re-mapping to the integer domain before output. This is especially useful in applications such as the RSA cryptosystem where modular exponentiation is the underlying operation. Efficient Montgomery multipliers for implementation on FPGA have previously been presented in [Daly and Marnane, 2002].

Kaliski's method requires a maximum of $(3m + 2)$ clock cycles to perform an m-bit Montgomery inversion, or $(4m + 2)$ cycles to perform a regular integer inversion. Montgomery inverters have previously been presented in [Gutub et al., 2002][Daly et al., 2003][Savas and Koc, 2000].

18.4 Modular Division

A new binary add-and-shift algorithm to perform modular division was presented recently by S.C. Shantz in [Shantz, 2001]. This provides an alternative to division by inversion followed by multiplication. The algorithm is given here in algorithm 1. It computes $U = \frac{y}{x} (\text{mod } M)$ in a maximum of $2(m - 1)$ clock cycles where m is the bitlength of the modulus, M.

Algorithm 1 : Modular Division

Input : y, $x \in [1, M-1]$ and M
Output : U, where $y = U * x \pmod{M}$

01. $A := x, B := M, U := y, V := 0$
02.   while $(A \neq B)$ do
03.   if $(A$ even) then
04. $A := A/2;$
05.   if $(U$ even) then $U := U/2$ else $U := (U + M)/2;$
06.   else if $(B$ even) then
07. $B := B/2;$
08.   if $(V$ even) then $V := V/2$ else $V := (V + M)/2;$
09.   else if $(A > B)$ then
10. $A := (A - B)/2;$
11. $U := (U - V);$
12.   if $(U < 0)$ then $U := (U + M);$
13.   if $(U$ even) then $U := U/2$ else $U := (U + M)/2;$
14.   else
15. $B := (B - A)/2;$
16. $V := (V - U);$
17.   if $(V < 0)$ then $V := (V + M);$
18.   if $(V$ even) then $V := V/2$ else $V := (V + M)/2;$
19.   end while
20.   return $U;$

18.5   Basic Division Architecture

As can be seen from algorithm 1, the divider makes decisions based on the parity and magnitude comparisons of m-bit registers A and B. To determine the parity of a number, only the Least Significant Bit (LSB) needs be examined (0 implies *even*, 1 implies *odd*). However, true magnitude comparisons can only be achieved through full m-bit subtractions, and thus introduce significant delay before decisions regarding the next computation can be made.

The first design presented here uses m-bit carry propagation adders to perform the additions/subtractions. At each iteration of the while loop in algorithm 1, U_{new} is assigned one of the following values, depending on the parity and relative magnitude of A and B: U, $\frac{U}{2}$, $\frac{(U+M)}{2}$, $\frac{(U-V)}{2}$, $\frac{(U-V+M)}{2}$ or $\frac{(U-V+2M)}{2}$.

The basis for the proposed designs is to calculate all 6 possible values concurrently, and use multiplexors to select the final value of U_{new}. This eliminates the time required to perform a full magnitude comparison of A and B before calculation of U and V can even commence. These architectures are illustrated in Fig. 18.3 and Fig. 18.4.

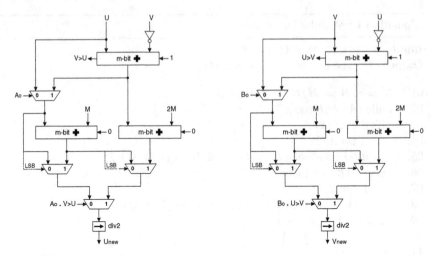

Figure 18.3. Determination of U_{new}.   *Figure 18.4.* Determination of V_{new}.

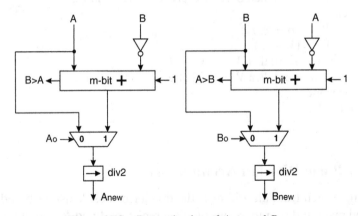

Figure 18.5. Determination of A_{new} and B_{new}.

The architecture for the determination of A_{new} and B_{new} is simpler as illustrated in Fig. 18.5. The U, V, A and B registers are also controlled by the parity and relative magnitudes of A and B, and are not clocked on every cycle.

18.6   Proposed Carry-Select Division Architecture

The clock speed is dependent on the bitlength of the divider since the carry chain of the adders contributes significantly to the overall critical path of the design. When the carry chain length exceeds the column height of the FPGA, the carry must be routed from the top of the column to the bottom of the next. This causes a significant decrease in the overall clock speed. The carry-select

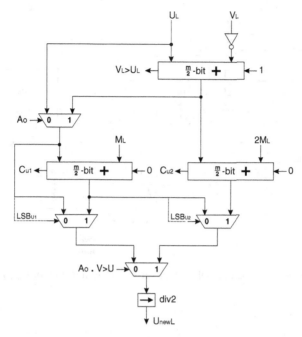

Figure 18.6. Carry-Select Architecture: Determination of U_{newL}.

design proposed here halves this adder carry chain at the expense of extra adders, multiplexors and control, but in doing so improves the performance of the divider.

This architecture is similar to a carry-select inverter design proposed in [Daly et al., 2003]. The values of $(U - V)$ and $(A - B)$ are determined by splitting the four m-bit registers U, V, A and B into eight $(\frac{m}{2})$-bit registers U_L, U_H, V_L, V_H, A_L, A_H, B_L and B_H. The values of $(U - V)_L$ and $(U - V)_H$ are then determined concurrently to produce $(U - V)$ as illustrated in Fig. 18.6 and Fig. 18.7.

When calculating $(U - V)$, if the least significant $(\frac{m}{2})$ bits of V are greater than the $(\frac{m}{2})$ least significant bits of U (i.e. $V_L > U_L$), then one extra bit must be "borrowed" from U_H to correctly determine the value of $(U - V)_H$. Therefore, the value of $(U - V)_H$ will actually be $(U_H - V_H - 1)$. It is observed that $(U_H - V_H - 1)$ is equivalent to $(\overline{V_H - U_H})$ in two's complement representation. Since the value of $(V_H - U_H)$ has been calculated in the determination of V_{new}, only a bitwise inverter is needed to produce $(U - V)_H$ as seen in Fig. 18.7. However, it is also possible that a carry from the addition of M_L or $2M_L$ will affect the final value of U_{newH}. Therefore carries of -1, 0 and 1 must be accounted for in the determination of U_{newH}. To allow for this, an extra $(\frac{m}{2})$-bit adder with a carry-in of 1 is required to calculate $(U_H - V_H + M_H + 1)$.

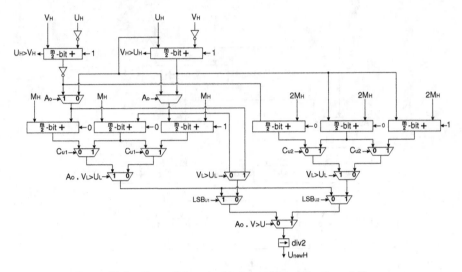

Figure 18.7. Carry-Select Architecture: Determination of U_{newH}.

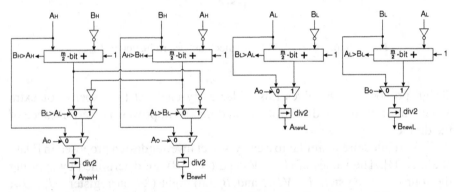

Figure 18.8. Carry-Select Architecture: Determination of A_{new} and B_{new}.

The determination of V_{new} is similar to that of U_{new}. The values of A_{new} and B_{new} are determined as illustrated in Fig. 18.8.

18.7 Results

Speed and area results for the two divider architectures are given in Tables 18.1 and 18.2. In each table the area and speed of each design are listed for 32, 64, 128 and 256-bit dividers, indicating the percentage increase in area and speed of the carry-select architecture over the basic architecture.

Table 18.1 lists the results targeted to the Xilinx VirtexE xcv2000e-6bg560 FPGA. This device has a maximum unbroken carry chain length of 160 bits (due to 80 Configurable Logic Blocks per column).

Table 18.1. Area and speed results for the two designs on xcv2000e-6bg560

Size	Design	Area (Slices)	% of xcv2000e	Area Increase	Max. Freq. (MHz)	Speed Increase
32-bit	*Basic*	561	2.9		63.16	
	Carry-Select	724	3.8	29.1%	57.25	−9.4%
64-bit	*Basic*	1,057	5.5		47.42	
	Carry-Select	1,351	7.0	27.8%	46.48	−2.0%
128-bit	*Basic*	1,992	10.4		30.90	
	Carry-Select	2,718	14.2	36.4%	35.18	13.9%
256-bit	*Basic*	4,353	22.7		19.85	
	Carry-Select	5,560	28.9	27.7%	26.23	32.1%

Table 18.2. Area and speed results for the two designs on xc2v1500-6bg575

Size	Design	Area (Slices)	% of xc2v1500	Area Increase	Max. Freq. (MHz)	Speed Increase
32-bit	*Basic*	551	7.2		103.39	
	Carry-Select	709	9.2	28.7%	91.62	−11.4%
64-bit	*Basic*	1,102	14.3		86.45	
	Carry-Select	1,398	18.2	26.9%	80.35	− 7.1%
128-bit	*Basic*	2,003	26.1		58.01	
	Carry-Select	2,622	34.1	30.9%	61.38	5.8%
256-bit	*Basic*	4,163	54.2		35.84	
	Carry-Select	5,268	68.6	26.5%	50.68	41.4%

Table 18.2 gives results targeted to the Xilinx Virtex2 xc2v1500-6bg575 FPGA. This device has a maximum unbroken carry chain length of 192 bits (due to 96 CLB's per column).

VHDL synthesis and place and route were performed on Xilinx ISE 5.1. The results are post place and route with a top level architecture to load the data in 32-bit words.

For 32-bit and 64-bit designs, the increased delay due to the extra level of multiplexors and control masks the reduction in the critical carry path of the adders. The delay introduced by the multiplexors is approximately 60 times that of a 1-bit carry propagation. This results in a decrease in speed for these designs.

For the 128-bit designs, the reduction of the carry chain from 128 bits to 64 bits in the carry-select design gives a slight speed improvement over the basic design. However, once the maximum carry chain length of the device has been exceeded, the basic architecture suffers a greater degradation in performance due to the routing of the carry from the top of one column to the bottom of the next. The 256-bit results show a much greater increase in speed for the carry-select design where the carry chain remains unbroken.

In all cases, the increase in area remains reasonably constant at approximately 30%.

Comparing these results to the inversion architecture results presented [Daly et al., 2003], it is observed that both inverter and divider have comparable operating clock frequencies.

The divider requires half the number of clock cycles to perform the operation, however it requires significantly more area. It is estimated that using this new architecture, division can be performed twice as fast as the alternative invert and multiply architecture.

18.8   Conclusions

Modular division is an important operation in elliptic curve cryptography. In this paper, two new FPGA architectures, based on a recently published division algorithm [Shantz, 2001] have been presented and implemented. The basic design computes all possible outcomes from each iteration and uses multiplexors to select the correct answer. This avoids the necessity to await the outcome of a full m-bit magnitude comparison before computation can begin. The second, carry-select divider design splits the critical carry chain into two, and again performs all calculations before the magnitude comparison has been completed. The 256-bit carry-select divider achieved a 40% increase in speed over the basic design at the cost of a 30% increase in area. The operating speed of the proposed divider is comparable to that of previously presented inverters on similar devices, but needs only half the number of clock cycles. Since division by inversion followed by multiplication requires many extra clock cycles, and hence more power, the proposed carry-select divider architecture seems better suited for implementation in an ECC processor.

Acknowledgments

This work is funded by the research innovation fund from Enterprise Ireland.

References

Blake, I., Seroussi, G., and Smart, N. (2000). *"Elliptic Curves in Cryptography"*. London Mathematical Society Lecture Note Series 265. Cambridge University Press.

Daly, A. and Marnane, W. (2002). "Efficient Architectures for Implementing Montgomery Modular Multiplication and RSA Modular Exponentiation on Reconfigurable Logic". 10th Intl Symposium on FPGA (FPGA 2002), pages 40–49.

Daly, A., Marnane, W., and Popovici, E. (2003). "Fast Modular Inversion in the Montgomery Domain on Reconfigurable Logic". *Irish Signals and Systems Conference ISSC2003*, pages 362–367.

Ernst, M., Jung, M., Madlener, F., Huss, S., and Blümel, R. (2002). "A Reconfigurable System on Chip Implementation for Elliptic Curve Cryptography over GF(2^n)". *Cryptographic Hardware and Embedded Systems—CHES 2002*, (LNCS 2523):381–398.

Gutub, A., Tenca, A. F., Savas, E., and Koc, C. K. (2002). "Scalable and unified hardware to compute Montgomery inverse in GF(p) and GF(2^n)". *Cryptographic Hardware and Embedded Systems - CHES 2002*, (LNCS 2523):484–499.

IEEE (2000). IEEE 1363/D13 Standard Specifications for Public Key Cryptography.

Kaliski, B. S. (1995). "The Montgomery Inverse and it's applications". *IEEE Trans. on Computers*, 44(8):1064–1065.

Kerins, T., Popovici, E., Marnane, W., and Fitzpatrick, P. (2002). "Fully Parameterizable Elliptic Curve Cryptography Processor over GF(2^m)". 12th *Intl Conference on Field-Programmable Logic and Applications FPL2002*, pages 750–759.

Koblitz, N. (1987). "Elliptic Curve Cryptosystems". *Math Comp*, 48:203–209.

Miller, V. S. (1985). "Use of Elliptic Curves in Cryptography". *Advances in Cryptography Crypto'85*, (218):417–426.

Orlando, G. and Paar, C. (2001). "A Scalable GF(p) Elliptic Curve Processor Architecture for Programmable Hardware". *Cryptographic Hardware and Embedded Systems - CHES 2001*, (LNCS 2162):348–363.

Savas, E. and Koc, C. K. (2000). "The Montgomery Modular Inverse—Revisited". *IEEE Trans. on Computers*, 49(7):763–766.

Shantz, S. C. (2001). "From Euclid's GCD to Montgomery Multiplication to the Great Divide". Technical Report TR-2001-95, Sun Microsystems Laboratories.

Chapter 19

A New Arithmetic Unit in $GF(2^M)$ for Reconfigurable Hardware Implementation

Chang Hoon Kim[1], Chun Pyo Hong[1], Soonhak Kwon[2] and
Yun Ki Kwon[2]

[1] Dept. of Computer and Information Engineering, Daegu University, Korea
chkim@dsp.daegu.ac.kr, cphong@daegu.ac.kr

[2] Dept. of Mathematics and Inst. of Basic Science, Sungkyunkwan University, Korea
shkwon@math.skku.ac.kr, drmath@galois10.skku.ac.kr

Abstract   In order to overcome the well-known drawback of reduced flexibility that is associated with traditional ASIC solutions, this paper proposes a new arithmetic unit (AU) in $GF(2^m)$ for reconfigurable hardware implementation such as FPGAs. The proposed AU performs both division and multiplication in $GF(2^m)$. These operations are at the heart of elliptic curve cryptosystems (ECC). Analysis shows that the proposed AU has significantly less area complexity and has roughly the same or lower latency compared with some related circuits. In addition, we show that the proposed architecture preserves a high clock rate for large m (up to 571), when it is implemented on Altera's EP2A70F1508C-7 FPGA device. Furthermore, since the new architecture does not restrict the choice of irreducible polynomials and has the features of regularity, modularity, and unidirectional data flow, it provides a high flexibility and scalability with respect to the field size m. Therefore, the proposed architecture is well suited for implementing both the division and multiplication units of ECC on FPGAs.

Keywords: Finite Field Division, Finite Field Multiplication, ECC, VLSI

19.1   Introduction

Information security has recently become an important subject due to the explosive growth of the Internet, mobile computing, and the migration of commerce practices to the electronic medium. The deployment of information security procedures requires the implementation of cryptosystems.

P. Lysaght and W. Rosenstiel (eds.),
New Algorithms, Architectures and Applications for Reconfigurable Computing, 231–249.
© 2005 Springer. Printed in the Netherlands.

Among these cryptosystems, ECC have recently gained a lot of attention in industry and academia. The main reason for the attractiveness of ECC is the fact that there is no sub-exponential algorithm known to solve the discrete logarithm problem on a properly chosen elliptic curve [1–3]. This means that significantly smaller parameters can be used in ECC relative to other competitive systems such as RSA and ElGamal with equivalent levels of security [2–3]. Some benefits of having smaller key sizes include faster computations, reductions in processing power, storage space, and bandwidth. Another advantage to be gained by using ECC is that each user may select a different elliptic curve, even though all users use the same underlying finite field. Consequently, all users require the same hardware for performing the field arithmetic, and the elliptic curve can be changed periodically for extra security [2].

Due to these many advantages of ECC, a number of software [4–5] and hardware [6–10] implementations have been documented. In those implementations, $GF(p)$, $GF(p^m)$, and $GF(2^m)$ have been used, where p is a prime. In particular, $GF(2^m)$, which is an m-dimensional extension field of $GF(2)$, is suitable for hardware implementation because there are no carry propagations in arithmetic operations. As is well known, software implementations can easily be achieved on a general-purpose microprocessor. However, they would be too slow for time critical applications and have weakness for side channel attacks such as timing and memory attacks [14]. Therefore, for performance as well as for physical security reasons, it is often required to realize ECC in hardware.

Although traditional ASIC solutions have significant advantages in area, speed, and power consumption, they have the well-known drawback of reduced flexibility compared to software solutions. Since modern security protocols are increasingly defined to be algorithm independent, a high degree of flexibility with respect to the cryptographic algorithms is desirable. A promising solution which combines high flexibility with the speed and physical security of traditional hardware is the implementation of cryptographic algorithms on re-configurable devices such as FPGAs [6–8], [17].

One of the most important design rules in FPGA implementation is the elimination of the broadcasting of global signals. This is because global signals in FPGAs are not simple wires but buses. i.e., they are connected by routing resources (switching matrices) having propagation delays [20], [23]. In general, since ECC require large field size (at least 163) to support sufficient security, when global signals are used, the critical path delay increases significantly [18]. Due to this problem, systolic array based designs, where each basic cell is connected with its neighboring cells through pipelining, are desirable to provide a higher clock rate and maximum throughput performance on fine-grained FPGAs [17–19].

In this paper, we propose a new AU, which performs both division and multiplication in $GF(2^m)$ and has a systolic architecture for implementing of ECC on FPGAs. The new design is achieved by using substructure sharing between the binary extended GCD algorithm [24] and the most significant bit (MSB)-first multiplication scheme [16]. When input data come in continuously, the proposed architecture produces division results at a rate of one per m clock cycles after an initial delay of $5m - 2$ in division mode and multiplication results at a rate of one per m clock cycles after an initial delay of $3m$ in multiplication mode respectively.

Analysis shows that the proposed AU has significantly less area complexity and has roughly the same or lower latency compared with some related systolic arrays for $GF(2^m)$. In addition, we show that the proposed architecture preserves a high clock rate for large m (up to 571), when it is implemented on Altera's EP2A70F1508C-7 FPGA device. Furthermore, since the new architecture does not restrict the choice of irreducible polynomials and has the features of regularity, modularity, and unidirectional data flow, it provides a high flexibility and scalability with respect to the field size m. Therefore, the proposed architecture is well suited for both division and multiplication unit of ECC implemented on fine-grained FPGAs.

19.2   Mathematical Background

19.2.1 $GF(2^m)$ Field Arithmetic for ECC

In ECC, computing kP is the most important arithmetic operation, where k is an integer and p is a point on elliptic curve. This operation can be computed using the addition of two points k times. Let $GF(2^m)$ be a finite field of characteristic 2. Then any non-supersingular elliptic curve E over $GF(2^m)$ can be written as

$$E : y^2 + xy = x^3 + a_1 x^2 + a_2 \tag{19.1}$$

where $a_1, a_2 \in GF(2^m)$, $a_2 \neq 0$, together with a special point called the point at infinity O. Let $P_1 = (x_1, y_1)$ and $P_2 = (x_2, y_2)$ be points in $E(GF(2^m))$ given in affine coordinates. Assume $P_1, P_2 \neq O$, and $P_1 \neq -P_2$. The sum $P_3 = (x_3, y_3) = P_1 + P_2$ is computed as follows [3]:

If $P_1 \neq P_2$	If $P_1 = P_2$ (called point doubling)
$\lambda = (y_1 + y_2)/(x_1 + x_2)$	$\lambda = y_1/x_1 + x_1$
$x_3 = \lambda^2 + \lambda + x_1 + x_2 + a_1$	$x_3 = \lambda^2 + \lambda + a_1$
$y_3 = (x_1 + x_3)\lambda + x_3 + y_1$	$y_3 = (x_1 + x_3)\lambda + x_3 + y_1$

As described above, the addition of two different elliptic curve points in $E(GF(2^m))$ requires one division, one multiplication, one squaring and eight additions in $GF(2^m)$. Doubling a point in $E(GF(2^m))$ requires one division, one multiplication, one squaring, and six additions respectively. Since the addition in $GF(2^m)$ is simply a bit-wise XOR operation, it can be implemented in fast and inexpensive ways. The squaring can be substituted by multiplication. Therefore, we will consider the multiplication and division in $GF(2^m)$. From the point addition formulae, it should be noted that any computation except for addition cannot be performed at the same time due to the data dependency. Therefore, sharing hardware between division and multiplication is more desirable than separate implementation of them.

19.2.2 $GF(2^m)$ **Field Arithmetic for ECC**

$GF(2^m)$ contains 2^m elements and is an extension field of $GF(2)$. All $GF(2^m)$ contain a zero element, a unit element, a primitive element, and have at least one irreducible polynomial $G(x) = x^m + g_{m-1}x^{m-1} + g_{m-2}x^{m-2} + \cdots + g_1x + g_0$ over $GF(2)$ associated with it. The primitive element α is a root of the irreducible polynomial $G(x)$ and generates all non-zero elements of $GF(2^m)$. The non-zero elements of $GF(2^m)$ can be represented as powers of the primitive element α, i.e., $GF(2^m) = \{1, \alpha^1, \alpha^2, \cdots, \alpha^{2^m-2}\}$. Since α is a root of the irreducible polynomial $G(x)$, i.e., $G(\alpha) = 0$, we have

$$\alpha^m = g_{m-1}\alpha^{m-1} + g_{m-2}\alpha^{m-2} + \cdots + g_1\alpha + g_0 \qquad (19.2)$$

Therefore, the elements of $GF(2^m)$ can also be represented as polynomials of α with a degree less than m by performing a modulo $G(\alpha)$ operation on α^t, where $0 \leq t \leq 2^m - 2$. This type of representation of the field elements is referred to as the standard basis (or the polynomial basis) representation. For example, the set $\{1, \alpha^1, \alpha^2, \cdots, \alpha^{m-1}\}$ forms the standard basis of $GF(2^m)$.

For the standard basis representation, the operations over $GF(2^m)$ are quite different from the usual binary arithmetic operations. The addition and subtraction of two field elements of $GF(2^m)$ are simply bit-wise XOR operation. For example, let $A(x)$ and $B(x)$ be two elements in $GF(2^m)$, then $A(x) + B(x) = \sum_{i=0}^{m-1}(a_i + b_i)x^i$, where the addition in the parenthesis indicates an XOR or modulo 2 addition operation. On the other hand, the multiplication and division operations are much more complicated. In particular, the division is the most time and area consuming operation. These operations can be performed by first multiplying $A(x)$ with $B(x)$ or $B(x)^{-1}$ and then taking modulo $G(x)$.

19.3   A New Dependence Graph for Both Division and Multiplication in $GF(2^m)$

19.3.1   Dependence Graph for Division in $GF(2^m)$

Let $A(x)$ and $B(x)$ be two elements in $GF(2^m)$, $G(x)$ be the irreducible polynomial used to generate the field $GF(2^m) \cong GF(2)[x]/G(x)$, and $P(x)$ be the result of the division $A(x)/B(x)$ mod $G(x)$. Then we may write

$$A(x) = a_{m-1}x^{m-1} + a_{m-2}x^{m-2} + \cdots + a_1x + a_0 \qquad (19.3)$$
$$B(x) = b_{m-1}x^{m-1} + b_{m-2}x^{m-2} + \cdots + b_1x + b_0 \qquad (19.4)$$

$$G(x) = x^m + g_{m-1}x^{m-1} + g_{m-2}x^{m-2} + \cdots + g_1x + g_0 \qquad (19.5)$$

$$P(x) = p_{m-1}x^{m-1} + p_{m-2}x^{m-2} + \cdots + p_1x + p_0 \qquad (19.6)$$

where the coefficients of each polynomial is binary digits 0 or 1. To compute the division $A(x)/B(x) \bmod G(x)$, Algorithm I can be used [24].

[Algorithm I] The Binary Extended GCD for Division in GF(2^m) [24]

Input: $G(x)$, $A(x)$, $B(x)$

Output: U has $P(x) = A(x)/B(x) \bmod G(x)$

Initialize: $R = B(x)$, $S = G = G(x)$, $U = A(x)$, $V = 0$,
$\qquad\qquad$ $count = 0$, $state = 0$

1. **for** $i=1$ **to** $2m-1$ **do**
2. $\quad$ **if** $state == 0$ **then**
3. $\qquad$ $count = count + 1$;
4. $\qquad$ **if** $r_0 == 1$ **then**
5. $\qquad\quad$ $(R, S) = (R + S, R);(U, V) = (U + V, U)$;
6. $\qquad\quad$ $state = 1$;
7. $\qquad$ **end if**
8. $\quad$ **else**
9. $\qquad$ $count = count-1$;
10. $\qquad$ **if** $r_0 == 1$ **then**
11. $\qquad\quad$ $(R, S) = (R + S, S); (U, V) = (U + V, V)$;
12. $\qquad$ **end if**
13. $\qquad$ **if** $count == 0$ **then**
14. $\qquad\quad$ $state = 0$;
15. $\qquad$ **end if**
16. $\quad$ **end if**
17. $\quad$ $R = R/x$;
18. $\quad$ $U = U/x$;
19. **end for**

Before deriving a new dependence graph (DG) corresponding to the Algorithm I for division in GF(2^m), we consider its main operations. Since R is a polynomial with degree of at most m and S is a polynomial with degree m, and U and V are polynomials with degree of at most $m - 1$, they can be expressed as follows:

$$R = r_mx^m + r_{m-1}x^{m-1} + \cdots + r_1x + r_0 \qquad (19.7)$$

$$S = s_mx^m + s_{m-1}x^{m-1} + \cdots + s_1x + s_0 \qquad (19.8)$$

$$U = u_{m-1}x^{m-1} + u_{m-2}x^{m-2} + \cdots + u_1x + u_0 \qquad (19.9)$$

$$V = v_{m-1}x^{m-1} + v_{m-2}x^{m-2} + \cdots + v_1x + v_0 \qquad (19.10)$$

As described in the Algorithm I, S and V are simple exchange operations with R and U respectively, depending on the value of $state$ and r_0. On the other

hand, R and U have two operational parts respectively. First, we consider the operations of R. Depending on the value of r_0, (R/x) or $((R+S)/x)$ is executed. Therefore, we can get the intermediate result of R as follows:
Let

$$R' = r'_m x^m + r'_{m-1} x^{m-1} + \cdots + r'_1 x + r'_0 = (R + r_0 S)/x \quad (19.11)$$

From r_0, (7), and (8), we get

$$r'_m = 0 \quad (19.12)$$

$$r'_{m-1} = r_0 s_m = r_0 s_m + 0 \quad (19.13)$$

$$r'_i = r_0 s_{i+1} + r_{i+1}, \, 0 \le i \le m - 2 \quad (19.14)$$

Second, we consider the operations of U. To get the intermediate result of U, we must compute the two operations of $U = (U + V)$ and $U = U/x$.
Let

$$U'' = u''_{m-1} x^{m-1} + \cdots + u''_1 x + u''_0 = U + V \quad (19.15)$$

From r_0, (9), and (10), we have

$$u''_i = r_0 v_i + u_i, \, 0 \le i \le m - 1 \quad (19.16)$$

Since $g_0 = 1$, we can rewrite (2) follows:

$$1 = (x^{m-1} + g_{m-1} x^{m-2} + g_{m-2} x^{m-3} + \cdots + g_2 x + g_1) x \quad (19.17)$$

From (17), we have

$$x^{-1} = x^{m-1} + g_{m-1} x^{m-2} + g_{m-2} x^{m-3} + \cdots + g_2 x + g_1 \quad (19.18)$$

Let

$$U''' = u'''_{m-1} x^{m-1} + \cdots + u'''_1 x + u'''_0 = U/x \quad (19.19)$$

By substituting (9) and (18) into (19), the following equations can be derived:

$$u'''_{m-1} = u_0 \quad (19.20)$$

$$u'''_i = u_{i+1} + u_0 g_{i+1}, \, 0 \le i \le m - 2 \quad (19.21)$$

Let

$$U' = u'_{m-1} x^{m-1} + \cdots + u'_1 x + u'_0 = U''/x \quad (19.22)$$

We can derive the following (23) and (24).

$$u'_{m-1} = r_0 v_0 + u_0 = (r_0 v_0 + u_0) g_m + r_0 0 + 0 \quad (19.23)$$

$$u'_i = (r_0 v_{i+1} + u_{i+1}) + (r_0 v_0 + u_0) g_{i+1}, \, 0 \le i \le m - 2 \quad (19.24)$$

In addition, the corresponding control functions of the Algorithm I are given as follows:

$$\text{Ctrl1} = (r_0 == 1) \tag{19.25}$$

$$\text{Ctrl2} = u_0 \text{ XOR } (v_0 \ \& \ r_0) \tag{19.26}$$

$$\text{Ctrl3} = (state == 0)\&(r_0 == 1) \tag{19.27}$$

$$count' = \begin{cases} count + 1, & \text{if } state == 0 \\ count - 1, & \text{if } state == 1 \end{cases} \tag{19.28}$$

$$state = \overline{state}, \ if \begin{cases} ((r_0 == 1) \ \& \ (state == 0)) \ or \\ ((count == 0) \ \& \ (state == 1)) \end{cases} \tag{19.29}$$

Based on the main operations and the control functions, we can derive a new DG for division in $GF(2^m)$ as shown in Fig. 19.1.

The DG corresponding to the Algorithm I consists of $(2m - 1)$ Type-1 cells and $(2m - 1) \times m$ Type-2 cells. In particular, we assumed $m = 3$ in the DG of Fig. 19.1, where the functions of two basic cells are depicted in Fig. 19.2 and Fig. 19.3 respectively. Note that, since r_m is always 0 and s_0 is always 1 in all

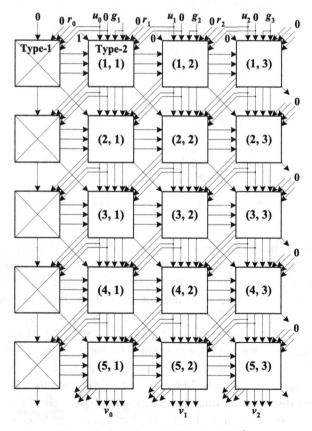

Figure 19.1. DG for division in $GF(2^3)$.

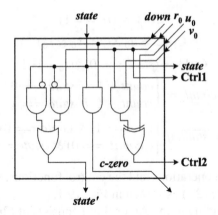

Figure 19.2. The circuit of Type-1 cell in Fig. 19.1.

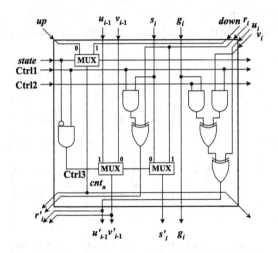

Figure 19.3. The circuit of Type-2 cell in Fig. 19.1.

the iterations of the Algorithm I, we do not need to process them. The input polynomials $R(x)$, $U(x)$, and $G(x)$ enter the DG from the top in parallel form. The i-th row of the array realizes the i-th iteration of the Algorithm I and the division result $V(x)$ emerge from the bottom row of the DG in parallel form after $2m - 1$ iterations. Before describing the functions of Type-1 and Type-2 cells, we consider the implementation of *count*. From the Algorithm I, since *count* increases to m, we can trace the value of *count* by putting $\log_2(m + 1)$-bit adders (subtractors) in each Type-1 cell. In addition, we can obtain a one-dimensional signal flow graph (SFG) array by projecting the DG in Fig. 19.1 along the east direction. In this case, since each basic cell of the SFG should contain one $\log_2(m + 1)$-bit adder (subtractor) circuit, the SFG array will have an area complexity of $O(\log_2 m)$. If m is very large such as ECC, it would be inefficient. To prevent such a problem, we adopt the method presented in [15]. In other words, we use m-bit bi-directional shift registers (BSR) instead of $\log_2(m+1)$-bit adders (subtractors) to compute *count*. The BSR is also used for multiplication, as will be explained in section 3.3.

In what follows, we add one 2-to-1 multipliexer, and *up* and *down* signals to each Type-2 cell. As shown in Fig. 19.1, in the first row, only the (1, 1)-th Type-2 cell receive 1, while the others receive 0 for *up*, and all the cells receive 0 for *down*. In the i-th iteration, m-bit BSR is shifted to the left or to the right according to *state*. When cnt_n in Fig. 19.3 is 1, it indicates that *count* becomes $n(1 \leq n \leq m)$. In addition, when *count* reduces to 0, *down* in Fig. 19.2 becomes 1. As a result, all the cnt_n of Type-2 cells in the same row become 0, and c-zero and *state* in Fig. 19.2 are updated to 1 and 0 respectively. This is the same condition as for the first computation.

With the *count* implementation results, we summarize the functions of Type-1 and Type-2 cells as follows:

(1) Type-1 cell: As depicted in Fig. 19.2, the Type-1 cell generates the control signals Ctrl1 and Ctrl2 for the present iteration, and updates *state* for the next iteration.

(2) Type-2 cell: the Type-2 cells in the i-th row generate the control signal Ctrl3 and perform the main operations of Algorithm I for the present iteration, and update *count* for the next iteration.

19.3.2 DG for MSB-first Multiplication in $GF(2^m)$

Let $A(x)$ and $B(x)$ be two elements in $GF(2^m)$, $G(x)$ be the irreducible polynomial, and $P(x)$ be the result of the multiplication $A(x)B(x) \bmod G(x)$. We can perform the multiplication using the following Algorithm II [16].

[Algorithm II] The MSB-first Multiplication Algorithm
in $GF(2^m)$ [16]
Input: $G(x)$, $A(x)$, $B(x)$
Output: $P(x) = A(x) B(x) \bmod G(x)$
1. $\quad p_k^{(0)} = 0$, for $0 \leq k \leq m-1$
2. $\quad p_{-1}^{(i)} = 0$, for $1 \leq i \leq m$
3. $\quad$**for** $i = 1$ **to** m **do**
4. $\quad\quad$**for** $k = m-1$ **down to** 0 **do**
5. $\quad\quad\quad p_k^{(i)} = p_{m-1}^{(i-1)} g_k + b_{m-1} a_k + p_{k-1}^{(i-1)},$
6. $\quad\quad$**end**
7. $\quad$**end**
8. $\quad P(x) = p^{(m)}(x)$

Based on the MSB-first multiplication algorithm in $GF(2^m)$, a DG can be derived as shown in Fig. 19.4 [16]. The DG corresponding to the multiplication algorithm consists of $m \times m$ basic cells. In particular, $m = 3$ in the DG of Fig. 19.4, and Fig. 19.5 represents the architecture of basic cells. The cells in the i-th row of the array perform the i-th iteration of the multiplication algorithm. The coefficients of the result $P(x)$ emerge from the bottom row of the array after m iterations.

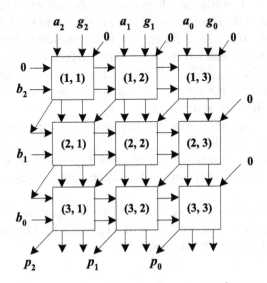

Figure 19.4. DG for multiplication in $GF(2^3)$ [16].

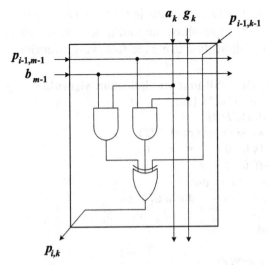

Figure 19.5. The basic cell of Fig. 19.4 [16].

19.3.3 A New DG for Both Division and Multiplication in $GF(2^m)$

By observing the DG in Fig. 19.1 and the DG in Fig. 19.4, we can find that the U operation of the division is identical with the p operation of the multiplication except for the input values. In addition, there are two differences between the DG for division and the DG for multiplication. First, in the DG for multiplication, the input polynomial $B(x)$ enters from the left, while, in the DG for division, all the input polynomials enter from the top. Second, the positions of the coefficients of each input polynomial are changed in two DGs. In this case, by modifying the circuit of each basic cell, both division and multiplication can be performed using the same hardware.

For the first case, by putting the m-bit BSR in each row of the DG for division, we can make the DG perform both division and multiplication depending on whether $B(x)$ enters from the left or from the top. In other words, after feeding $B(x)$ into m-bit BSR, it is enough to shift it to the left until the multiplication is finished. For the second case, we can use the same hardware by permuting the coefficients of each input polynomial. In summary, the DG in Fig. 19.1 with appropriate modification can perform both division and multiplication. The resulting DG is shown in Fig. 19.6, and the corresponding Modified Type-1 and Modified Type-2 cells in Fig. 19.6 are shown in Fig. 19.7 and Fig. 19.8 respectively.

As described in Fig. 19.6, the DG consists of m Modified Type-1 cells, $m \times m$ Modified Type-2 cells, $m - 1$ Type-1 cells, and $(m - 1) \times m$ Type-2 cells. The polynomials $A(x)$, $B(x)$, and $G(x)$ enter the DG for multiplication, and the polynomials $R(x)$, $U(x)$, and $G(x)$ enter the DG for division. The multiplication result $P(x)$ emerges from the bottom row of Part-A after m iterations, and the division result $V(x)$ emerge from the bottom row of Part-B after $2m - 1$ iterations. In Fig. 19.6, Part-A is used for both division and multiplication and Part-B is only used for division. The modification procedures from Type-1, 2 to Modified Type-1, 2 are summarized as follows:

(a) Modification of Type-1 cell: As described in Fig. 19.7, we have added a *mult/div* signal, a 2-to-1 AND gate (numbered by 1) and a 2-to-1 multiplexer when compared to a Type-1 cell. For multiplication mode, *mult/div* is set to 0 and for division mode, *mult/div* is set to 1. As a result, for multiplication, b_{m-i}/Ctrl1 has b_{m-i} and p_{m-1}/Ctrl2 has p_{m-1} respectively. For division, it has the same function as the Type-1 cell in Fig. 19.2.

(b) Modification of Type-2 cell: As described in Fig. 19.8, we have added a *mult/div* signal, a 2-to-1 AND gate (numbered by 1), a 2-to-1 OR gate (numbered by 2) and a 2-to-1 multiplexer when compared to a Type-2 cell. In Fig. 19.8, since Ctrl3 generates 0 for multiplication mode, a_{m-i}/v_{i-1} is always selected. For multiplication mode, the m-bit BSR is only shifted to the left direction due to the OR gate numbered by 2. In addition, the multiplexer numbered by 4 in

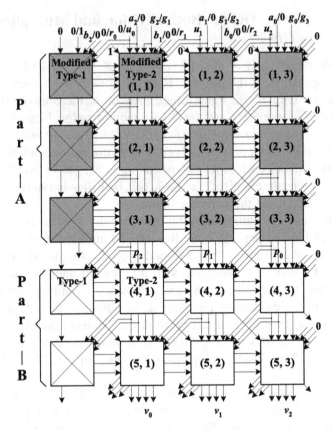

Figure 19.6. New DG for both division and multiplication in $GF(2^3)$.

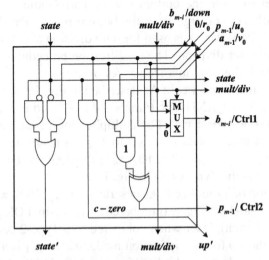

Figure 19.7. The circuit of Modified Type-1 cell in Fig. 19.6.

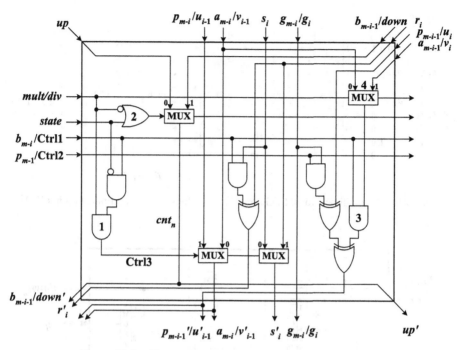

Figure 19.8. The circuit of Modified Type-2 cell in Fig. 19.6.

Fig. 19.8 is also added due to the fact that, for multiplication mode, the AND gate number by 3 must receive a_{m-i}/v_{m-1} instead of a_{m-i-1}/v_i. As a result, when *mult/div* is set to 0, the Modified Type-2 cell in Fig. 19.8 performs as the basic cell in Fig. 19.5. In addition, when *mult/div* is set to 1, it performs as the Type-2 cell in Fig. 19.3.

19.4 A New AU for Both Division and Multiplication in $GF(2^m)$

By projecting the DG in Fig. 19.6 along the east direction according to the projection procedure [21], we derive a one-dimensional SFG array as shown in Fig. 19.9, where the circuit of each processing element (PE-A and PE-B) is described in Fig. 19.10 and Fig. 19.11 respectively. In Fig. 19.9, " • " denotes a 1-bit 1-cycle delay element. The SFG array of Fig. 19.9 is controlled by a sequence $011 \cdots 11$ of length m. As described in Fig. 19.9, two different data set are feed into the array depending on whether division or multiplication is selected.

As described in Fig. 19.10, the PE-A contains the circuitry of Modified Type-1 (Fig. 19.7) and Type-2 (Fig. 19.8) cells. In addition, as described in Fig. 19.11, the PE-B contains the circuitry of Type-1 (Fig. 19.2) and Type-2 (Fig. 19.3)

244

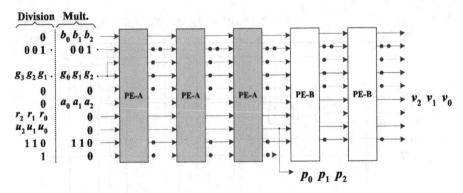

Figure 19.9. A one-dimensional SFG array for both division and multiplication in $GF(2^3)$.

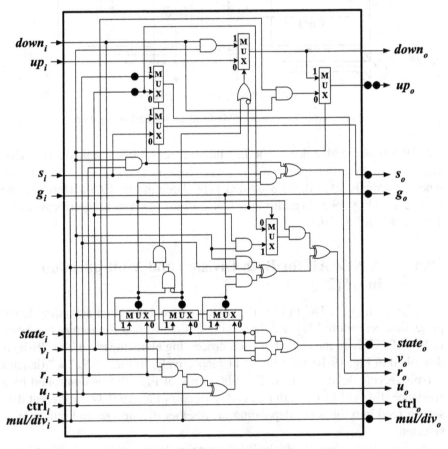

Figure 19.10. The circuit of PE-A in Fig. 19.9.

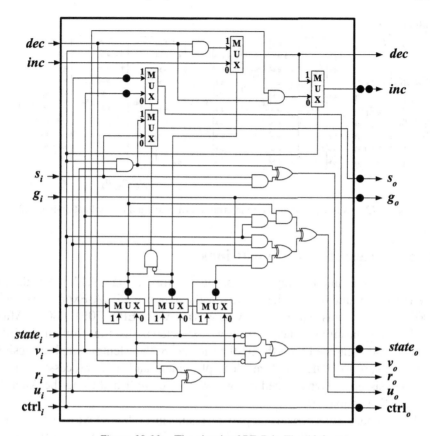

Figure 19.11. The circuit of PE-B in Fig. 19.9.

tcells. Since two control signals Ctrl1 and Ctrl2, and the *state* of the i-th iteration must be broadcast to all the Modified Type-1 and Type-2 cells in the i-th row of the DG, three 2-to-1 multiplexers and three 1-bit latches are added to each PE-A and PE-B. This result is depicted in Fig. 19.10 and Fig. 19.11. Four 2-input AND gates are also added to each PE-A and PE-B due to the fact that four signals (i.e., $b_{m-i-1}/down$, r_i, p_{m-i-1}/u_i, and a_{m-i-1}/v_i in Fig. 19.6) must be fed to each row of the DG from the rightmost cell. When the control signal is at logic 1, these AND gates generate four zeros.

The SFG array in Fig. 19.9 can be easily retimed by using the cut-set systolization techniques [21] to derive a serial-in serial-out systolic array and the resulting structure is shown in Fig. 19.12. When the input data come in continuously, this array can produce division results at a rate of one per m clock cycles after an initial delay of $5m - 2$ with the least significant coefficient first in division mode. It can produce multiplication results at a rate of one per m clock cycles after an initial delay of $3m$ with the most significant coefficient first in multiplication mode.

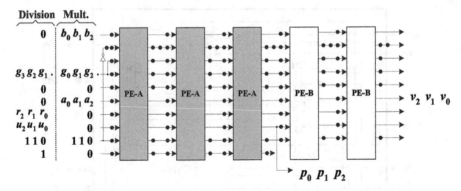

Figure 19.12. A new bit-serial systolic array for division and multiplication in $GF(2^3)$.

19.5 Results and Conclusions

To verify the functionality of the proposed array in Fig. 19.12, it was developed in VHDL and was synthesized using the Synopsis FPGA-Express (version 2000,11-FE3.5 tools. The target device was an Altera EP2A70F1508C-7. After synthesizing the circuits successfully, we extracted netlist files from the FPGA-Express and simulated with the netlist files using the Mentor Graphics design view software (VHDL-ChipSim). The placement and route process and timing analysis of the synthesized designs were accomplished using the Altera's Quartus II (version 2.0).

We summarize theoretical comparison results in Table 19.1 with some related systolic arrays having the same I/O format. It should be mentioned that there is no circuit for both division and multiplication in $GF(2^m)$ at this moment to the authors knowledge. In addition, FPGA implementation results of our architecture are given in Table 19.2. As described in Table 19.1, all the bit-serial approaches including the proposed array achieve the same time complexity of $O(m)$. However, the proposed systolic array reduces the area complexity from $O(m^2)$ or $O(m \cdot \log_2 m)$ to $O(m)$, and has lower maximum cell delay and latency than the architecture in [13]. Although the circuits in [11–12] have lower maximum cell delay than the proposed array, they cannot be applied to ECC due to their high area complexity of $O(m^2)$. In addition, since the proposed array performs both division and multiplication, we do not need additional hardware components described in Table 19.3 to achieve multiplication.

From Table 19.2, it is noted that the proposed architecture preserves a high clock rate for large field size m because there is no broadcasting of global signals. As a reference, 571 in Table 19.2 is the largest field size recommended by NIST [22]. Furthermore, since the proposed architecture does not restrict the choice of irreducible polynomials and has the features of regularity, modularity, and unidirectional data flow, it provides a high flexibility and scalability with

Table 19.1. Comparison with bit-serial systolic arrays

	[11]	[12]	[13]	Proposed AU
Throughput (1/cycles)	$1/(2m-1)$	$1/m$	$1/m$	Division : $1/m$ Mult. : $1/m$
atency (cycles)	$7m-3$	$5m-1$	$8m-1$	Division : $5m-2$ Mult. : $3m$
Maximum cell delay	$T_{AND2}+T_{XOR2}$ $+T_{MUX2}$	$T_{AND2}+T_{XOR2}$ $+T_{MUX2}$	$2T_{AND2}+2T_{XOR2}$ $+2T_{MUX2}$	$2T_{AND2}+T_{XOR2}$ $+T_{MUX2}$
Basic components and their numbers	AND$_2$: $3m^2+3m-2$ XOR$_2$: $1.5m^2+1.5m-$ 1 MUX$_2$: $3m^2+m-2$ Latch : $6m^2+8m-4$	AND$_2$: $2m-1$ OR$_2$: $3m$ XOR$_2$: $0.5m^2+1.5m-1$ MUX$_2$: $1.5m^2+4.5m-2$ Latch : $2.5m^2+14.5m-6$	Inverter : $2m$ AND$_2$: $26m$ XOR$_2$: $11m$ MUX$_2$: $35m+2$ FILO (m-bit) : 4 Latch : $46m +$ $4m \cdot \log_2(m + 1)$ zero-check $(\log_2(m + 1)$-bit) : $2m$ adder/subtractor $(\log_2(m + 1)$-bit) : $2m$	Inverter : $7m-2$ AND$_2$: $26m-$ 12 OR$_2$: $3m-1$ XOR$_2$: $8m-4$ MUX$_2$: $15m-7$ Latch : $44m-24$
Operation	Division	Division	Division	Division and Multiplication
Area	$O(m^2)$	$O(m^2)$	$O(m \cdot \log_2 m)$	$O(m)$

AND$_i$: i-input AND gate
XOR$_i$: i-input XOR gate
OR$_i$: i-input OR gate
MUX$_i$: i-to-1 multiplexer
T_{ANDi}: the propagation delay through one AND$_i$ gate
T_{XORi}: the propagation delay through one XOR$_i$ gate
T_{MUXi}: the propagation delay through one MUX$_i$ gate

Table 19.2. FPGA Implementation results of the propose AU

	163	233	409	571
# of LE	9920	14209	24926	34950
Clock(MHz)	138.87	127.99	141.8	121.11
Chip Utilization(%)	14.76	21.21	37.09	52.01

LE consists of one 4-to-1 LUT, one flip-flop, fast carry logic, and programmable multiplexers [23]

respect to the field size m. All these advantages of our design suggest that if ECC is implemented on FPGAs to overcome the well-known drawback of ASIC, we can obtain maximum throughput performance with minimum hardware requirement by using the proposed AU.

248

Table 19.3. Area-time complexity of the bit-serial systolic array for multiplication in $GF(2^m)$ [16]

Througput	Latency	Maximum cell delay	Basic components and their numbers
$1/m$	$3m$	$T_{AND2} + 2T_{XOR2}$	$AND_2 : 3m$, $XOR_2 : 2m$ $MUX_2 : 2m$, Latch $: 10m$

Acknowledgments

This work was supported by grant No. R05-2003-000-11573-0 from the Basic Research Program of the Korea Science & Engineering Foundation

References

[1] IEEE P1363, *Standard Specifications for Publickey Cryptography*, 2000.

[2] A. Menezes, *Elliptic Curve Public Key Cryptosystems*, Kluwer Academic Publishers, 1993.

[3] I. F. Blake, G. Seroussi, and N. P. Smart, *Elliptic Curves in Cryptography*, Cambridge University Press, 1999.

[4] D. Hankerson, J. L. Hernandez, and A. Menezes, "Implementation of Elliptic Curve Cryptography Over Binary Fields," *CHES 2000*, LNCS 1965, Springer-Verlag, 2000.

[5] D. Bailey and C. Paar, "Efficient Arithmetic in Finite Field Extensions with Application in Elliptic Curve Cryptography," *J. of Cryptology*, vol. 14, no. 3, pp. 153–176, 2001.

[6] L. Gao, S. Shrivastava and G. E. Solbelman, "Elliptic Curve Scalar Multiplier Design Using FPGAs," *CHES 2000*, LNCS 1717, Springer-Verlag, 1999.

[7] G. Orlando and C. Parr, "A High-Performance Reconfigurable Elliptic Curve Processor for $GF(2^m)$," *CHES 2000*, LNCS 1965, Springer-Verlag, 2000.

[8] M. Bednara, M. Daldrup, J. von zur Gathen, J. Shokrollahi, and J. Teich, "Reconfigurable Implementation of Elliptic Curve Crypto Algorithms," *Proc. of the International Parallel and Distributed Processing Symposium (IPDPS'02)*, pp. 157–164, 2002.

[9] G. B. Agnew, R. C. Mullin, and S. A. Vanstone, "An Implementation for Elliptic Curve Cryptosystems Over F_2^{155}," *IEEE J. Selected Areas in Comm.*, vol. 11, no. 5, pp. 804–813, June 1993.

[10] M.A. Hasan and A.G. Wassal, "VLSI Algorithms, Architectures, and Implementation of a Versatile $GF(2^m)$ Processor", *IEEE Trans. Computers*, vol. 49, no. 10, pp. 1064–1073, Oct. 2000.

[11] C.-L. Wang and J. -L. Lin, "A Systolic Architecture for Computing Inverses and Divisions in Finite Fields $GF(2^m)$," *IEEE Trans. Computers.*, vol. 42, no. 9, pp. 1141–1146, Sep. 1993.

[12] M.A. Hasan and V.K. Bhargava, "Bit-Level Systolic Divider and Multiplier for Finite Fields $GF(2^m)$," *IEEE Trans. Computers*, vol. 41, no. 8, pp. 972–980, Aug. 1992.

[13] J.-H. Guo and C.-L. Wang, "Systolic Array Implementation of Euclid's Algorithm for Inversion and Division in $GF(2^m)$," *IEEE Trans. Computers.*, vol. 47, no. 10, pp. 1161–1167, Oct. 1998.

[14] J.R. Goodman, "Energy Scalable Reconfigurable Cryptographic Hardware for Portable Applications," PhD thesis, MIT, 2000.

[15] J.-H. Guo and C.-L. Wang, "Bit-serial Systolic Array Implementation of Euclid's Algorithm for Inversion and Division in $GF(2^m)$", *Proc. 1997 Int. Symp. VLSI Tech., Systems and Applications*, pp. 113–117, 1997.

[16] C. L. Wang and J. L. Lin, "Systolic Array Implementation of Multipliers for Finite Field $GF(2^m)$," *IEEE Trans. Circuits and Syst.*, vol. 38, no. 7, pp. 796–800, July 1991.

[17] T. Blum and C. Paar, "High Radix Montgomery Modular Exponentiation on Reconfigurable Hardware", *IEEE Trans. Computers.*, vol. 50, no. 7, pp. 759–764, July 2001.

[18] S.D. Han, C.H. Kim, and C.P. Hong, "Characteristic Analysis of Modular Multiplier for $GF(2^m)$," *Proc. of IEEK Summer Conference 2002*, vol. 25, no. 1, pp. 277–280, 2002.

[19] R. Tessier and W. Burleson, "Reconfigurable Computing for Digital Signal Processing: A Survey", *J. VLSI Signal Processing*, vol. 28, no. 1, pp. 7–27, May 1998.

[20] K. Compton and S. Hauck, "Reconfigurable Computing: A Survey of Systems and Software", *ACM Computing Surveys*, vol. 34, no. 2, pp. 171–210, June 2002.

[21] S. Y. Kung, *VLSI Array Processors*, Englewood Cliffs, NJ: Prentice Hall, 1988.

[22] NIST, Recommended elliptic curves for federal government use, May 1999. http://csrc.nist.gov/ encryption.

[23] Altera, APEX$^{TM II}$ Programable Logic Device Family Data Sheet, Aug. 2000. http://www.altera.com/ literature/lit-ap2.html.

[24] C.H. Kim and C.P. Hong, "High Speed Division Architecture for $GF(2^m)$", *Electronics Letters*, vol. 38, no. 15, pp. 835–836, July 2002.

Chapter 20

Performance Analysis of SHACAL-1 Encryption Hardware Architectures

Máire McLoone[1], J.V. McCanny[2]

[1] *Institute for Electronics, Communications and Information Technology (ECIT), Queen's University Belfast, Stranmillis Road, Belfast BT9 5AH, Northern Ireland*
maire.mcloone@ee.qub.ac.uk

[2] *Institute for Electronics, Communications and Information Technology (ECIT), Queen's University Belfast, Stranmillis Road, Belfast BT9 5AH, Northern Ireland*
j.mccanny@ee.qub.ac.uk

Abstract   Very high-speed and low-area hardware architectures of the SHACAL-1 encryption algorithm are presented in this paper. The SHACAL algorithm was a submission to the New European Schemes for Signatures, Integrity and Encryption (NESSIE) project and it is based on the SHA-1 hash algorithm. Sub-pipelined SHACAL-1 encryption and decryption architectures are described and when implemented on Virtex-II XC2V4000 FPGA devices, run at a throughput of 23 Gbps. In addition, fully pipelined and iterative architectures of the algorithm are presented. The SHACAL-1 decryption algorithm is derived and also presented in the paper, since it was not provided in the submission to NESSIE.

Keywords: NESSIE, SHACAL

20.1   Introduction

New European Schemes for Signatures, Integrity and Encryption (NESSIE) was a three-year research project, which formed part of the Information Societies Technology (IST) programme run by the European Commission. The main aim of NESSIE is to present strong cryptographic primitives covering confidentiality, data integrity and authentication, which have been open to a rigorous evaluation and cryptanalytical process. Therefore, submissions not only included private key block ciphers, as with the Advanced Encryption Standard

251

P. Lysaght and W. Rosenstiel (eds.),
New Algorithms, Architectures and Applications for Reconfigurable Computing, 251–264.
© 2005 *Springer. Printed in the Netherlands.*

(AES) project [1], but also hash algorithms, digital signatures, stream ciphers and public key algorithms. The first call for submissions was in March 2000 and resulted in 42 submissions. The first phase of the project involved conducting a security and performance evaluation of all the submitted algorithms. The performance evaluation hoped to achieve performance estimates for software, hardware and smartcard implementations of each algorithm. 24 algorithms were selected for further scrutiny in the second phase, which began in July 2001. The National Institute for Standards and Technology's (NIST) Triple-DES and Rijndael private key algorithms and SHA-1 and SHA-2 hash algorithms were also considered for evaluation. The following 7 block cipher algorithms were chosen for further investigation in phase two of the NESSIE project: IDEA, SHACAL, SAFER++, MISTY1, Khazad, RC6 and Camellia. The final selection of the NESSIE cryptographic primitives took place on 27 February 2003. MISTY1, Camellia, SHACAL-2 and Rijndael are the block ciphers chosen to be included in the NESSIE portfolio of cryptographic primitives. Overall, 12 of the original submissions are included along with 5 existing standard algorithms.

Currently, the fastest known FPGA implementations of the NESSIE finalist algorithms are as follows: Pan et al [2] estimate that their IDEA architecture can achieve a throughput of 6 Gbps on a Virtex-II XC2V1000 FPGA device. Rouvroy et al [3] report a MISTY1 algorithm implementation which runs at 19.4 Gbps on Virtex-II XC2V2000 device. Standaert et al.'s [4] Khazad architecture runs at 9.5 Gbps on the XCV1000 FPGA. A pipelined Camellia implementation by Ichikawa et al [5] on a Virtex-E XCV1000E device performs at 6.7 Gbps. The fastest FPGA implementation of the RC6 algorithm is the 15.2 Gbps implementation by Beuchat [6] on the Virtex-II XC2V3000 device. Finally, an iterative architecture of the SAFER++ algorithm by Ichikawa et al [7] achieves a data-rate of 403 Mbps on a Virtex-E XCV1000E FPGA. In this paper, a performance evaluation of SHACAL-1 hardware architectures is provided which improves on results previously published by the authors [8]. Prior to this research, only published work on SHACAL-1 software implementations was available.

Section 2 of the paper provides a description of the SHACAL-1 algorithm. Section 3 outlines the SHACAL-1 architectures for hardware implementation. Performance results are given in section 4 and conclusions are provided in section 5.

20.2   A Description of the SHACAL-1 Algorithm

The SHACAL [9] algorithm, developed by Helena Handschuh and David Naccache of Gemplus, is based on the NIST's SHA hash algorithm. SHACAL is defined as a variable block and key length family of ciphers. There are two versions specified in the submission: SHACAL-1, which is derived from SHA-1 and SHACAL-2, which is derived from SHA-256. However, other versions can easily be derived from the SHA-384 and SHA-512 hash algorithms. Only the

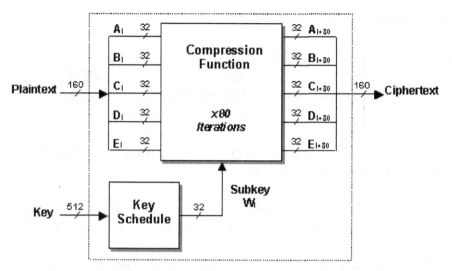

Figure 20.1. Outline of SHACAL-1 Encryption Algorithm.

SHACAL-1 algorithm is considered for implementation in this paper. However, the SHACAL-1 architecture described can be easily adapted for the other variants.

SHACAL-1 operates on a 160-bit data block utilising a 512-bit key. The key, k, can vary in length within the range, $128 \leq k \leq 512$. If $k < 512$, it is appended with zeros to a length of 512-bits. SHACAL-1 encryption, outlined in Fig. 20.1 is performed by splitting the 160-bit plaintext into five 32-bit words, A, B, C, D and E. Next, 80 iterations of the compression function are performed. The resulting A, B, C, D and E values are concatenated to form the ciphertext. The compression function is defined as follows:

$$A_{i+1} = ROT_{LEFT-5}(A_i) + F_i(B_i, C_i, D_i) + E_i + Cnst_i + W_i$$
$$B_{i+1} = A_i$$
$$C_{i+1} = ROT_{LEFT-30}(B_i) \qquad (20.1)$$
$$D_{i+1} = C_i$$
$$E_{i+1} = D_i$$

where, W_i are the subkeys generated from the key schedule and the ROT_{LEFT-n} function is defined as a 32-bit word rotated to the left by n positions. The constants, $Cnst_i$, in hexadecimal, are,

$$Cnst_i = 5a827999 \ \ 0 \leq i \leq 19$$
$$Cnst_i = 6ed9eba1 \ \ 20 \leq i \leq 39$$
$$Cnst_i = 8f1bbcdc \ \ 40 \leq i \leq 59 \qquad (20.2)$$
$$Cnst_i = ca62c1d6 \ \ 60 \leq i \leq 79$$

and the function $F_i(x, y, z)$ is defined as,

$$F_i(x, y, z) = \begin{cases} (x \text{ } AND \text{ } y)OR(\overline{x} \text{ } AND \text{ } z) & 0 \le i \le 19 \\ x \oplus y \oplus z & 20 \le i \le 39 \\ (x \text{ } AND \text{ } y)OR(x \text{ } AND \text{ } z)OR(y \text{ } AND \text{ } z) & 40 \le i \le 59 \\ x \oplus y \oplus z & 60 \le i \le 79 \end{cases} \quad (20.3)$$

In the SHACAL-1 key schedule, the 512-bit input key is expanded to form eighty 32-bit subkeys, W_i, such that,

$$W_i = \begin{cases} Key_i & 0 \le i \le 15 \\ ROT_{LEFT-1}(W_{i-3} \oplus W_{i-8} \oplus W_{i-14} \oplus W_{i-16}) & 16 \le i \le 79 \end{cases} \quad (20.4)$$

The first 16 subkeys are formed by splitting the input key into sixteen 32-bit values.

20.2.1 SHACAL-1 Decryption

SHACAL-1 decryption is not defined in the submission. Therefore, it has been derived and is presented here. It requires an inverse compression function and an inverse key schedule.

The inverse compression function is as follows:

$$\begin{aligned} A_{i+1} &= B_i \\ B_{i+1} &= ROT_{RIGHT-30}(C_i) \\ C_{i+1} &= D_i \\ D_{i+1} &= E_i \\ E_{i+1} = A_i - [ROT_{LEFT-5}(B_i) &+ InvF_i(ROT_{RIGHT-30}(C_i), D_i, E_i) \\ &+ InvCnst_i + InvW_i] \end{aligned} \quad (20.5)$$

where $InvW_i$ are the inverse subkeys generated from the inverse key schedule and the $ROT_{RIGHT-n}$ function is defined as a 32-bit word rotated to the right by n positions. The constants, $InvCnst_i$, in hexadecimal, are,

$$\begin{aligned} InvCnst_i &= ca62c1d6 & 0 \le i \le 19 \\ InvCnst_i &= 8f1bbcdc & 20 \le i \le 39 \\ InvCnst_i &= 6ed9eba1 & 40 \le i \le 59 \\ InvCnst_i &= 5a827999 & 60 \le i \le 79 \end{aligned} \quad (20.6)$$

and the function $InvF_i(x, y, z)$ is defined as,

$$
InvF_i(x, y, z) = \begin{cases} x \oplus y \oplus z & 0 \le i \le 19 \\ (xANDy)OR(xANDz) & \\ OR(yANDz) & 20 \le i \le 39 \\ x \oplus y \oplus z & 40 \le i \le 59 \\ (xANDy)OR(\overline{x}ANDz) & 60 \le i \le 79 \end{cases} \tag{20.7}
$$

The subkeys generated from the key schedule during encryption are used in reverse order when decrypting data. Therefore, it is necessary to wait for all of the subkeys to be generated before beginning decryption. However, an inverse key schedule can be utilised to generate the subkeys in the order that they are required for decryption. This inverse key schedule is defined as,

$$
InvW_i = \begin{cases} InvKey_i & 0 \le i \le 15 \\ ROT_{RIGHT-1}(InvW_{i-16}) \oplus InvW_{i-13} & \\ \oplus InvW_{i-8} \oplus InvW_{i-2} & 16 \le i \le 79 \end{cases} \tag{20.8}
$$

The *InvKey* is created by concatenating the final 16 subkeys generated during encryption, such that,

$$
InvKey_i = W_{64}, W_{65}, W_{66}, W_{67}, W_{68}, W_{69},
$$
$$
W_{70}, W_{71}, W_{72}, W_{73}, W_{74,75}, W_{76}, W_{77}, W_{78} \tag{20.9}
$$

20.3   SHACAL-1 Hardware Architectures

The SHACAL-1 algorithm is derived from the SHA hash algorithm. Therefore, the design of a SHACAL-1 hardware architecture can be derived from the design of a SHA hash algorithm architecture. Previous efficient implementations [10–12] of the SHA-1 and SHA-2 hash algorithms have utilised a shift register design approach. Thus, this methodology has also been used in the iterative, fully and sub- pipelined SHACAL-1 architectures described here.

20.3.1   Iterative SHACAL-1 Architectures

In the iterative encryption and decryption architectures data is passed through the compression or inverse compression function component eighty times. The initial five data blocks are generated from the input block split into five 32-bit blocks. The outputs of the function, A to E form the inputs on consecutive clock cycles. After 80 iterations, the A to E outputs are concatenated to form the 160-bit plaintext/ciphertext.

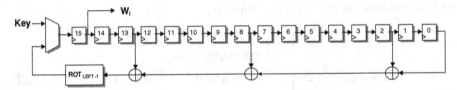

Figure 20.2. SHACAL-1 Key Schedule Design.

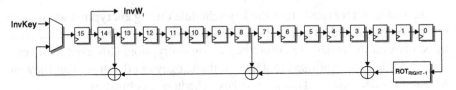

Figure 20.3. SHACAL-1 Inverse Key Schedule Design.

The main components in the iterative, fully and sub- pipelined architectures are the compression function and the key schedule. The key schedule is implemented using a 16-stage shift register design, as illustrated in Fig. 20.2.

The input 512-bit key is loaded into the registers in 32-bit blocks over 16 clock cycles. On the next clock cycle, the value of register 15 is updated with the result of Eqn. 20.4. An initial delay of 16 clock cycles is avoided by taking the values, W_i from the output of register 15 and not from the output of register 0. The subkeys, W_i, are generated as they are required by the compression function.

The inverse key schedule is designed in a similar manner and is shown in Fig. 20.3.

Typically, for decryption, the sender of the ciphertext will send the receiver the original key used to encrypt the message. Hence, the receiver will have to generate all eighty 32-bit subkeys before commencing decryption. However, this can be avoided if the sender of the ciphertext sends the receiver the final sixteen 32-bit subkeys that were created during encryption of the message as a 512-bit inverse key. Now, the receiver can immediately begin to decrypt the ciphertext, since the subkeys required for decryption can be generated as they are required using this inverse key.

The design of the SHACAL-1 compression function is depicted in Fig. 20.4. The design requires 5 registers to store the continually updating values of A, B, C, D and E. The values in registers B, C and D are operated on by one of four different functions every 20 iterations, as given in Eqn. 20.3.

The critical path of the overall SHACAL design occurs in the compression function in the calculation of,

$$A_{i+1} = ROT_{LEFT-5}(A_i) + F_i(B_i, C_i, D_i) + E_i + Cnst_i + W_i \quad (20.10)$$

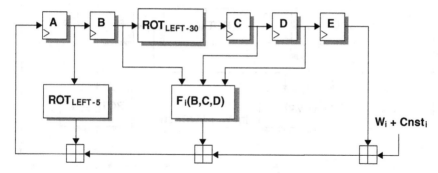

Figure 20.4. SHACAL-1 Compression Function Design.

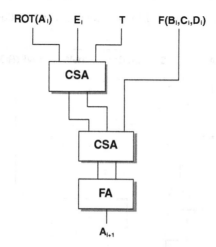

Figure 20.5. CSA Implementation in Compression Function.

In the architectures described here, this critical path is reduced in 2 ways. Firstly, the addition, $T = Cnst_i + W_i$, is performed on the previous clock cycle and thus, is removed from the critical path. Also, Carry-Save-Adders (CSAs) are utilised. With CSAs, the carry propagation is avoided until the final addition. Therefore, Eqn. 20.10 is implemented using two CSAs and one Full Adder (FA) rather than three FAs, as shown in Fig. 20.5. Since the CSAs involve 32-bit additions, this implementation is faster than an implementation using only FAs [13].

The inverse compression function, outlined in Fig. 20.6, contains a subtraction. Hence, the throughput of the SHACAL-1 decryption architecture will be slower than that of the encryption architecture.

During decryption the critical path of the overall SHACAL design occurs in the inverse compression function, where,

$$E_{i+1} = A_i - [ROT_{LEFT-5}(B_i) + InvF_i(ROT_{RIGHT-30}(C_i), D_i, E_i)$$
$$+ InvCnst_i + InvW_i] \tag{20.11}$$

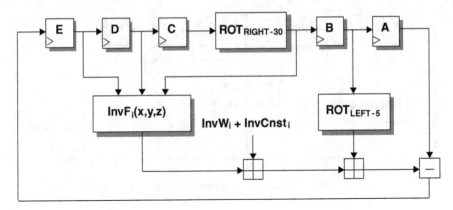

Figure 20.6. SHACAL-1 Inverse Compression Function Design.

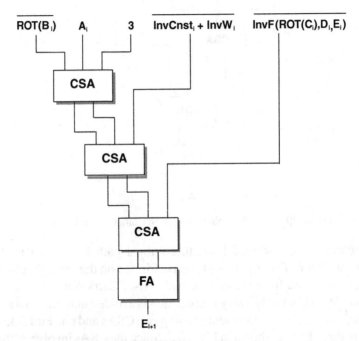

Figure 20.7. CSA Implementation in Inverse Compression Function.

Once again, it can be reduced by performing the addition, $InvCnst_i + InvW_i$, on the previous clock cycle. Eqn. 20.11 is manipulated for a CSA implementation by using two's complement logic, such that,

$$E_{i+1} = A_i + \overline{ROT_{LEFT-5}(B_i)} + 1$$
$$+ \overline{Inv\,F_i(ROT_{RIGHT-30}(C_i),\ D_i,\ E_i)} + 1 + \overline{Inv\,Cnst_i + InvW_i} + 1 \quad (20.12)$$

Thus, CSAs can be used to improve the overall efficiency of the decryption architecture, as illustrated in Fig. 20.7.

20.3.2 Fully and Sub-Pipelined SHACAL-1 Architectures

In the pipelined SHACAL-1 encryption architecture, the compression function and key schedule are designed as for the iterative architecture. However, the eighty compression function iterations are fully unrolled and registers placed between each component, as depicted in Fig. 20.8. It is assumed that the same key is used throughout a data transfer session. Every twenty compression function components contain a different function according to Eqn. 20.3. New plaintext blocks can be accepted on every clock cycle and after an initial delay of 81 clock cycles, the corresponding ciphertext blocks will appear on consecutive clock cycles. This leads to a very high-speed design. Further increases in speed can be obtained by sub-pipelining each compression function. After implementation using CSAs, the critical path of the encryption architecture lies in the path from C_i to A_{i+1}. In order to reduce this path, a further set of registers can be placed between the CSAs, as shown in Fig. 20.9. The initial latency is now 162 clock cycles before ciphertext blocks are output.

The fully and sub- pipelined decryption architectures are designed in a similar manner using the inverse compression function and inverse key schedule. After implementation using CSAs, the critical path of the decryption architecture lies in the path from E_i to E_{i+1}. A further set of registers can be placed between the CSAs, as shown in Fig. 20.10, to reduce this new critical path.

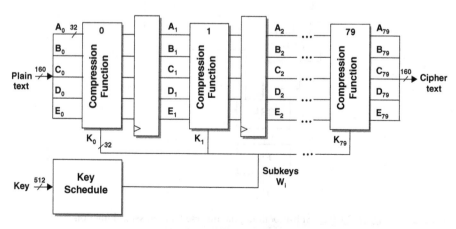

Figure 20.8. SHACAL-1 Fully Pipelined Architecture.

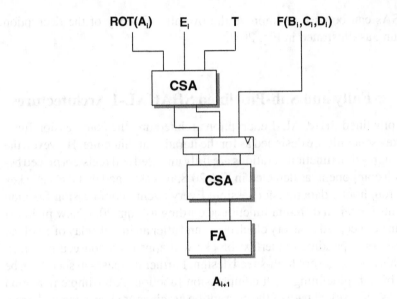

Figure 20.9. Sub-Pipelining the Compression Function.

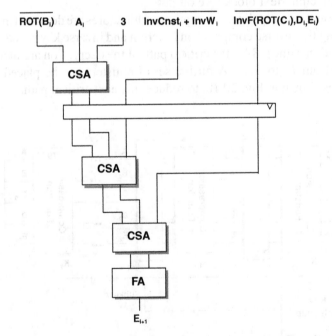

Figure 20.10. Sub-Pipelining the Inverse Compression Function.

20.4   Performance Evaluation

To provide a hardware performance evaluation for SHACAL-1, the hardware architectures described in this paper are implemented on Xilinx FPGA devices for demonstration purposes. The designs were simulated using Modelsim and synthesised using Synplify Pro v7.2 and Xilinx Foundation Series 5.1i software. They were verified using the SHACAL-1 test vectors provided in the submission to NESSIE.

The iterative encryption architecture implemented on the Virtex-II XC2V500 device runs at a clock speed of 110 MHz and hence, achieves a throughput of 215 Mbps. The design utilises just 994 CLB slices. The iterative decryption architecture implemented on the same device has a data-rate of 186 Mbps and requires 1039 slices. The decryption design is slower since its critical path contains a subtraction. The iterative architectures result in highly-compact yet efficient implementations. In both designs the 160-bit plaintext/ciphertext blocks and 512-bit key are input and output in 32-bit blocks. This implies a lower IOB count and less routing, thus, the designs can be implemented efficiently on smaller FPGA devices.

The fully and sub- pipelined architectures result in very high-speed designs. The performance results for these architectures are outlined in Table 20.1.

The sub-pipelined architectures achieve higher speeds than the fully pipelined architectures at the of additional area. However, the overall efficiency of the designs is not affected and the sub-pipelined architecture can be implemented on the same device. Additional sub-pipelining of the compression function could be carried out to further increase throughput. However, the additional area would require that the design is implemented on a larger Virtex-II device.

Even higher data-rates are attainable if the architectures are implemented on ASIC technology. Table 20.2 provides a summary of published work on

Table 20.1.   Performance Results of SHACAL-1 Fully and Sub-Pipelined Architectures

Architecture	Device	Speed (MHz)	Area (slices)	Throughput (Mbps)	Efficiency (Mbps/slices)
Fully-pipelined Encryption	XC2V4000	107	13729	17184	1.25
Fully-pipelined Decryption	XC2V4000	104	15241	16668	1.09
Sub-pipelined Encryption	XC2V4000	145	19733	23242	1.2
Sub-pipelined Decryption	XC2V4000	135	23038	23118	1

fast, pipelined hardware implementations of the NESSIE block cipher finalists on FPGA devices. The SAFER++ algorithm is not included in the table since an iterative design with a throughput of 403 Mbps [7] is the fastest implementation published to date. For comparison purposes, the table includes the performance metrics of a sub-pipelined single-chip implementation [14] of the Rijndael algorithm, which is the fastest FPGA implementation of this algorithm reported to date. Overall, the SHACAL-1 algorithm provides the fastest pipelined implementation. However, it is difficult to compare the performance of the algorithms as they are implemented on different devices. The closest in speed to the SHACAL-1 implementation is the 20 Gbps Rijndael design by Saggese et al [14] on a Virtex-E device. This Rijndael design requires a large number of memory blocks and as such, the Virtex-E FPGA is the most suitable device for implementation since it comprises a memory-rich architecture. The SHACAL-1 algorithm requires no memory blocks and therefore, the Virtex-II FPGA is targeted, as opposed to the Virtex-E device.

Although it is the fastest, the pipelined design is also the largest in area. High-speed, yet lower area implementations are possible by unrolling the algorithm by a lower number of compression function components, while still adopting pipelining. The SHACAL specification provided performance metrics for a software implementation of SHACAL-1 on an 800 MHz Pentium III processor. Encryption was achieved at a data-rate of 52 Mbps, decryption at 55 Mbps and key setup at 180 Mbps [9]. Therefore, even the iterative hardware design outlined in this paper is 4 times faster than their software implementation.

The efficiency of the algorithm implementations is also given in Table 20.2. Efficiency calculations are not provided for implementations that have been designed specifically to a device, and thus, include features such as multipliers and BRAM components. The MISTY1 implementation is the most efficient pipelined design while the SHACAL-1 implementation on the Virtex-II device is the next most efficient.

20.5   Conclusions

Throughout the NESSIE project, there has been no performance evaluation provided for hardware implementations of the SHACAL algorithm. In this paper, the SHACAL-1 algorithm is studied and both low-area iterative and high-speed pipelined architectures are described. The results presented in the paper improve on performance metrics previously published by the authors [8]. The SHACAL-1 architectures are implemented on Xilinx FPGA devices for demonstration purposes, but can readily be implemented on other FPGA or ASIC technologies. The iterative architectures are highly compact, yet efficient, when implemented on Virtex-II devices. A very high-speed 23 Gbps design is achieved when the sub-pipelined SHACAL-1 architecture is implemented on

Table 20.2. Summary of NESSIE Algorithm Hardware Implementations.

Authors	Algorithm	Device	Area	Throughput (Mbps)	Efficiency (Mbps/slices)
Authors	SHACAL-1	XC2V4000	19733 slices	23242	1.2
Pan *et al* [2]	IDEA	XC2V1000	4221 slices 34 multipliers	6080	–
Rouvroy *et al* [3]	MISTY1	XC2V2000	6322 slices	19392	3.07
Standaert *et al* [4]	Khazad	XCV1000	8800 slices	9472	1.08
Ichikawa *et al* [5]	Camellia	XCV1000E	9692 slices	6750	0.7
Beuchat[6]	RC6	XC2V3000	8554 slices 80 multipliers	15200	–
Saggese *et al* [14]	Rijndael	XCV2000E	5810 slices 100 BRAM	20224	–

the Virtex-II XC2V4000 device. SHACAL-1 is the fastest algorithm in hardware when compared to pipelined hardware implementations of the other NESSIE block cipher finalists. The other SHACAL algorithm version specified in the submission, SHACAL-2, operates on a 256-bit data block utilising a 512-bit key. The SHACAL-1 architectures described in this paper can be easily adapted for SHACAL-2. The authors have also conducted work on SHACAL-2 iterative and pipelined architectures, which achieve even higher speeds than SHACAL-1 architectures. The SHACAL-2 iterative design runs at a speed of 432 Mbps while a fully pipelined architecture achieves a throughput of 26 Gbps [15].

The SHACAL-2 algorithm was selected as one of four block ciphers to be included in the NESSIE portfolio of cryptographic primitives. Overall, the SHACAL algorithm has proven to be very secure with a large security margin. Since it is derived from the SHA hash algorithm, it has already undergone much cryptanalysis. It performs well when implemented in software [16] and as is evident from this paper, very high-speed hardware implementations are also possible. Since security and performance were the two main criteria in the NESSIE selection process, the SHACAL algorithm is an ideal candidate for selection.

References

[1] US NIST Advanced Encryption Standard, (January, 2004) http://csrc. nist.gov/encryption/aes/

[2] Z. Pan, S. Venkateswaran, S.T. Gurumani, and B.E. Wells, Exploiting Fine-Grain Parallelism present within the International Data Encryption Algorithm using a Xilinx FPGA, 16th International Conference on Parallel and Distributed Computing Systems (PDCS-2003), (Nevada, US, 2003).

[3] G. Rouvroy, F.X. Standaert, J.J. Quisquater, J.D. Legat, Efficient FPGA Implementation of Block Cipher MISTY1, The Reconfigurable Architecture Workshop (RAW 2003), (Nice, France, 2003).

[4] F.X. Standaert, G. Rouvroy, Efficient FPGA Implementation of Block Ciphers Khazad and MISTY1, 3rd NESSIE Workshop, http://www.di.ens.fr/wwwgrecc/NESSIE3/, (Germany, 2002).

[5] T. Ichikawa, T. Sorimachi, T. Kasuya, M. Matsui, On the criteria of hardware evaluation of block ciphers(1), Technical report of IEICE, ISEC2001-53, (2001).

[6] J.L. Beuchat, High Throughput Implementations of the RC6 Block Cipher Using Virtex-E and Virtex-II Devices, INRIA Research Report, http://perso.ens-lyon.fr/jean-luc.beuchat/publications.html, (July, 2002).

[7] T. Ichikawa, T. Sorimachi, T. Kasuya, On Hardware Implementation of Block Ciphers Selected at the NESSIE Project Phase 1, 3rd NESSIE Workshop, http://www.di.ens.fr/wwwgrecc/NESSIE3/, (Germany, November 2002).

[8] M. McLoone, J.V. McCanny, Very High Speed 17 Gbps SHACAL Encryption Architecture, 13th International Conference on Field Programmable Logic and Applications—FPL 2003, (Portugal, September 2003).

[9] H. Handschuh, D. Naccache, SHACAL, 1st NESSIE Workshop, http://www.cosic.esat.kuleuven.ac.be/nessie/workshop/, (Belgium, November, 2000).

[10] K.K. Ting, S.C.L. Yuen, K.H. Lee, P.H.W. Leong, An FPGA based SHA-256 Processor, 12th International Conference on Field Programmable Logic and Applications—FPL 2002, (France, September, 2002).

[11] T. Grembowski, R. Lien, K. Gaj, N. Nguyen, P. Bellows, J. Flidr, T. Lehman, B. Schott, Comparative Analysis of the Hardware Implementations of Hash Functions SHA-1 and SHA-512, Information Security Conference", (October, 2002).

[12] M. McLoone, J.V. McCanny, Efficient Single-Chip Implementation of SHA-384 and SHA-512, IEEE International Conference on Field-Programmable Technology—FPT 2002), (Hong Kong, Dec 2002).

[13] T. Kim, W. Jao, S. Tjiang, Circuit Optimization Using Carry-Save-Adder Cells, IEEE Transactions on Computer-Aided Design of Integrated Circuits and Systems, Vol. 17, No. 10, (October, 1998).

[14] G.P. Saggese, A. Mazzeo, N. Mazzocca, A.G.M. Strollo, An FPGA-Based Performance Analysis of the Unrolling, Tiling, and Pipelining of the AES Algorithm, 13th International Conference on Field Programmable Logic and Applications—FPL 2003, (Portugal, September, 2003).

[15] M. McLoone, Hardware Performance Analysis of the SHACAL-2 Encryption Algorithm, submitted to IEE Proceedings—Circuits, Devices and Systems, (July, 2003).

[16] NESSIE, Performance of Optimized Implementations of the NESSIE Primitives, http://www.cosic.esat.kuleuven.ac.be/nessie/deliverables/D21-v2.pdf, (February, 2003).

Chapter 21

Security Aspects of FPGAs in Cryptographic Applications

Thomas Wollinger and Christof Paar*

Chair for Communication Security (COSY)
Ruhr-Universität Bochum, Germany
{wollinger, cpaar}@crypto.rub.de

Abstract   This contribution provides a state-of-the-art description of security issues on FPGAs from a system perspective. We consider the potential security problems of FPGAs and propose some countermeasure for the existing drawbacks of FPGAs. Even though there have been many contributions dealing with the algorithmic aspects of cryptographic schemes implemented on FPGAs, this contribution is one of the few investigations of system and security aspects.

Keywords: cryptography, security, attacks, reconfigurable hardware, FPGA, cryptographic applications, reverse engineering

21.1   Introduction and Motivation

In recent years, FPGAs manufacturers have come closer to filling the performance gap between FPGAs and ASICs, enabling them, not only to serve as fast prototyping tools but also to become active players as components in systems. Reconfigurable hardware devices seem to combine the advantages of software and hardware implementations. Furthermore, there are potential advantages of reconfigurable hardware in cryptographic applications: algorithm agility, algorithm upload, architecture efficiency, resource efficiency, algorithm modification, throughput and cost efficiency.

The choice of the implementation platform of a digital system is driven by many criteria and is heavily dependent on the application area. In addition to

---

* This research was partially sponsored by the German Federal Office for Information Security (BSI).

P. Lysaght and W. Rosenstiel (eds.),
New Algorithms, Architectures and Applications for Reconfigurable Computing, 265–278.

the aspects of algorithm and system speed and costs there are crypto-specific ones: physical security (e.g., against key recovery and algorithm manipulation); flexibility (regarding algorithm parameter, keys, and the algorithm itself); power consumption (absolute usage and prevention of power analysis attacks); and other side channel leakages.

The remainder of this chapter is organized as follows: We devote the first part of this chapter to studying FPGAs from a system security perspective by describing some possible attacks (Section 21.2). In the second part we present possible countermeasures against the introduced attacks (Section 21.3). We end this contribution with some conclusions.

21.2 Shortcomings of FPGAs for Cryptographic Applications

This section summarizes security problems produced by attacks against given FPGA implementations. First we would like to state what the possible goals of such attacks are.

21.2.1 Why does Someone Wants to Attack FPGAs?

The most common threat against an implementation of a cryptographic algorithm is to learn a confidential cryptographic key, that is, either a symmetric key or the private key of an asymmetric algorithm. Given that the algorithms applied are publicly known in most commercial applications, knowledge of the key enables the attacker to decrypt future (assuming the attack has not been detected and countermeasures have not been taken) and, often more harming, past communications which had been encrypted. Another threat is the one-to-one copy, or "cloning", of a cryptographic algorithm *together* with its key. In some cases it can be enough to run the cloned application in decryption mode to decipher past and future communications. In other cases, execution of a certain cryptographic operation with a presumingly secret key is in most applications the sole criteria which authenticates a communicating party. An attacker who can perform the same function can masquerade as the attacked communicating party. Yet another threat is presented by applications where the cryptographic algorithms are proprietary. Even though such an approach is not widespread, it is standard practice in applications such as pay-TV and in government communications. In such scenarios it is already interesting for an attacker to reverse-engineer the encryption algorithm itself. The associated key might later be recovered by other methods (e.g., bribery or classical cryptanalysis.) The discussion above assumes mostly that an attacker has physical access to the encryption device. Whether that is the case or not depends heavily on the application. However, we believe that in many scenarios such access can be assumed, either through outsiders or through dishonest insiders.

In the following we discuss the vulnerabilities of modern FPGAs to such attacks. In areas where no attacks on FPGAs have been published, we have tried to extrapolate from attacks on other hardware platforms, mainly memory cell and chip cards.

21.2.2   Description of the Black Box Attack

The classical method to reverse engineer a chip is the so-called Black Box attack. The attacker inputs all possible combinations, while saving the corresponding outputs. The intruder is then able to extract the inner logic of the FPGA with the help of the Karnaugh map or algorithms that simplify the resulting tables. This attack is only feasible if a small FPGA with explicit inputs and outputs is attacked and a lot of processor power is available. The reverse engineering effort grows and it will become less feasible as the size and complexity of the FPGA increases. The cost of the attack, furthermore, rises with the usage of state machines, LFSRs (Linear Feedback Shift Registers), integrated storage, and, if pins can be used, as input and output [Dipert, 2000].

21.2.3   Cloning of SRAM FPGAs

The security implications that arise in a system that uses SRAM FPGAs are obvious, if the configuration data is stored unprotected in the system but external to the FPGA. In a standard scenario, the configuration data is stored externally in nonvolatile memory (e.g., PROM) and is transmitted to the FPGA at power up in order to configure the FPGA. An attacker could easily eavesdrop on the transmission and get the configuration file. This attack is therefore feasible for large organizations as well as for those with low budgets and modest sophistication.

21.2.4   Description of the Readback Attack

Readback is a feature that is provided for most FPGA families. This feature allows one to read a configuration out of the FPGA for easy debugging. An overview of the attack is given in [Dipert, 2000]. The idea of the attack is to read the configuration of the FPGA through the JTAG or programming interface in order to obtain secret information (e.g. keys or a proprietary algorithm). The readback functionality can be prevented with a security bit. In some FPGA families, more than one bit is used to disable different features, e.g., the JTAG boundary. In [Aplan et al., 1999], the idea of using a security antifuse to prevent readout of information is patented.

However, it is conceivable, that an attacker can overcome these countermeasures in FPGAs with fault injection. This kind of attack was first introduced in [Boneh et al., 1997]. The authors showed how to break public-key algorithms, such as the RSA and Rabin signature schemes, by exploiting hardware faults. Furthermore, they give a high-level description of transient faults, latent faults, and induced faults. This publication, was followed by [Biham and Shamir, 1997], where the authors introduced differential fault analysis, which can potentially be applied against all symmetric algorithms in the open literature. Meanwhile there have been many publications that show different techniques to insert faults, e.g., electro magnetic radiation [Quisquater and Samyde, 2001], infrared laser [Ajluni, 1995], or even a flash light [Skorobogatov and Anderson, 2002]. It seems very likely that these attacks can be easily applied to FPGAs, since they are not especially targeted to ASICs. Therefore, one is able to deactivate security bits and/or the countermeasures, resulting in the ability to read out the configuration of the FPGA [Kessner, 2000, Dipert, 2000].

Despite these attacks Actel Corporation [Actel Corporation, 2002] claims that after the programming phase, the cells of their FPGAs cannot be read at all. On the other hand Xilinx offers the users the software tool JBits [Guccione and Levi, 2003], which provides an API to access the bitstream information and allows dynamic reconfiguration for Xilinx Virtex FPGAs. JBits allows a simplified and automated access to specific parts of the bitstream, resulting in a extra advantage for the attacker who performs a readback attack.

21.2.5   Reverse-Engineering of the Bitstreams

The attacks described so far output the bitstream of the FPGA design. In order to get the design of proprietary algorithms or the secret keys, one has to reverse-engineer the bitstream. The condition to launch the attack is not only that the attacker has to be in possession of the bitstream, but furthermore the bitstream has to be in the clear, meaning it is not encrypted.

FPGA manufactures claim, that the security of the bitstream relies on the disclosure of the layout of the configuration data. This information will only be made available if a non-disclosure agreement is signed, which is, from a cryptographic point of view, an extremely insecure situation. This security-by-obscurity approach was broken at least ten years ago when the CAD software company NEOCad reverse-engineered a Xilinx FPGA. NEOCad was able to reconstruct the necessary information about look-up tables, connections, and storage elements [Seamann, 2000]. Hence, NEOCad was able to produce design software without signing non-disclosure agreements with the FPGA manufacturer. Even though a big effort has to be made to reverse engineer the bitstream, for large organizations it is quite feasible. In terms of government organizations as attackers, it is also possible that they will get the information of the design methodology directly from the vendors or companies that signed NDAs.

21.2.6 Description of Side Channel Attacks

Any physical implementation of a cryptographic system might provide a side channel that leaks unwanted information. Examples for side channels include in particular: power consumption, timing behavior, and electromagnet radiation. Obviously, FPGA implementations are also vulnerable to these attacks. In [Kocher et al., 1999] two practical attacks, Simple Power Analysis (SPA) and Differential Power Analysis (DPA) were introduced. The power consumption of the device while performing a cryptographic operation was analyzed in order to find the secret keys from a tamper resistant device. The main idea of DPA is to detect regions in the power consumption of a device which are correlated with the secret key. Moreover, in some cases little or no information about the target implementation is required. Since their introduction, there has been a lot of work improving the original power attacks (see, e.g., relevant articles in [Kaliski, Jr. et al., 2003]. There seems to be very little work at the time of writing addressing the feasibility of actual side channel attacks against FPGAs. Very recently the first experimental results of simple power analysis on an ECC implementation on an FPGA have been presented in [Örs et al., 2003] and on RSA and DES implementations in [Standaert et al., 2003]. Somewhat related was the work presented in [Shang et al., 2002] which concludes that 60% of the power consumption in a XILINX Virtex-II FPGA is due to the interconnects and 14% and 16% is due to clocking and logic, respectively. These figures would seem to imply that an SPA type attack would be harder to implement on an FPGA than on an ASIC. However, the results presented in [Standaert et al., 2003, Örs et al., 2003] show that SPA attacks are feasible on FPGAs and that they can be realized in practice.

21.2.7 Description of Physical Attacks

The aim of a physical attack is to investigate the chip design in order to get information about proprietary algorithms or to determine the secret keys by probing points inside the chip. Hence, this attack targets parts of the FPGA, which are not available through the normal I/O pins. This can potentially be achieved through visual inspections and by using tools such as optical microscopes and mechanical probes. However, FPGAs are becoming so complex that only with advanced methods, such as Focused Ion Beam (FIB) systems, one can launch such an attack. To our knowledge, there are no countermeasures to protect FPGAs against this form of physical threat. In the following, we will try to analyze the effort needed to physically attack FPGAs manufactured with different underlying technologies.

SRAM FPGAs. Unfortunately, there are no publications available that accomplished a physical attack against SRAM FPGAs. This kind of attack is only treated very superficially in a few articles, e.g. [Richard, 1998]. In the related area of SRAM memory, however there has been a lot of effort by academia and industry to exploit this kind of attack [Gutmann, 1996, Gutmann, 2001, Anderson and Kuhn, 1997, Williams et al., 1996, Schroder, 1998, Soden and Anderson, 1993, Kommerling and Kuhn, 1999]. Due to the similarities in structure of the SRAM memory cell and the internal structure of the SRAM FPGA, it is most likely that the attacks can be employed in this setting.

Contrary to common wisdom, SRAM memory cells do not entirely loose their contents when power is cut. The reason for these effects are rooted in the physical properties of semiconductors (see [Gutmann, 2001] for more details). The physical changes are caused mainly by three effects: electromigration, hot carriers, and ionic contamination.

Electromigration implies a high current density, that relocates metal atoms in the opposite direction of the current flow. Electromigration results in voids at the negative electrode and hillocks and whiskers at the positive electrode. The result is that electrons with very high energy are able to overcome the $Si - SiO_2$ potential barrier and accelerate into the gate oxide. These are called hot carrier electrons and it can be days before they become neutralized [Gutmann, 2001]. Ionic contamination is triggered by the sodium ions present in the material which are used during the semiconductor manufacturing and packaging processes. Electrical fields and high temperature enable movement towards the silicon/sillicon-dioxide interfaces, resulting in a changes in threshold voltages.

In the published literature one can find several different techniques to determine the changes in device operations. Most publications agree that devices can be altered, if 1) threshold voltage has changed by 100 mV or 2) there is a 10% change in transconductance, voltage or current. An extreme case of data recovery, was described in [Anderson and Kuhn, 1997]. The authors were able to extract a DES master key from a module used by a bank, without any special techniques or equipment on power-up. The reason for this was that the key was stored in the same SRAM cells over a long period of time. Hence, the key was "burned" into the memory cells and the key values were retained even after switching off the device.

"I_{DDQ} testing" is a widely used method for testing integrated circuits and is based on the analysis of the current usage of a device. The idea is to execute a set of test vectors until a given location is reached, at which point the device current is measured. Hot carrier effects, cell charge, and transitions between different states can then be detected as abnormal I_{DDQ} characteristics [Gutmann, 2001, Williams et al., 1996]. In [Schroder, 1998], the authors use the substrate current, the gate current, and the current in the drain-substrate diode of a MOSFET to determine the level and duration of stress applied.

When it becomes necessary to access internal portions of a device, there are also alternative techniques available to do so, as described in [Soden and Anderson, 1993]. Possibilities include the use of the scan path that the IC manufacturers insert for test purposes or techniques like bond pad probing [Gutmann, 2001].

When it becomes necessary to use access points that are not provided by the manufacturer, the layers of the chip have to be removed. Mechanical probing with tungsten wire with a radius of $0, 1 - 0, 2$ μm is the traditional way to discover the needed information. These probes provide gigahertz bandwidth with $100fF$ capacitance and $1M\Omega$ resistance.

Due to the complex structure and the multi-layer production of chips, mechanical testing is not sufficient enough. Focused Ion Beam (FIB) workstations can expose buried conductors and deposit new probe points. The functionality is similar to an electron microscope and one can inspect structures down to 5nm [Kommerling and Kuhn, 1999]. Use of an electron-beam tester (EBT) is another measurement method. An EBT is a special electron microscope that is able to speed primary electrons up to 2.5 kV at 5nA. EBT measures the energy and amount of secondary electrons that are reflected.

From the discussion above, it seems likely that a physical attack against SRAM FPGAs can be launched successfully, assuming that the described techniques can be transferred. However, the physical attacks are quite costly and having the structure and the size of state-of-the-art FPGA in mind, the attack will probably only be possible for large organizations, for example intelligence services.

Antifuse FPGAs.   To discuss physical attacks against antifuse (AF) FPGAs, one has to first understand the programming process and the structure of the cells. The basic structure of an AF node is a thin insulating layer (smaller than 1 μm^2) between conductors that is programmed by applying a voltage. After applying the voltage, the insulator becomes a low-resistance conductor and there exists a connection (diameter about 100nm) between the conductors. The programming function is permanent and the low-impedance state will persist indefinitely.

In order to be able to detect the existence or non-existence of the connection one has to remove layer after layer, or/and use cross-sectioning. Unfortunately, no details have been published regarding this type of attack. In [Dipert, 2000], the author states that a lot of trial-and-error is necessary to find the configuration of one cell and that it is likely that the rest of the chip will be destroyed, while analyzing one cell. The main problem with this analysis is that the isolation layer is much smaller than the whole AF cell. One study estimates that about 800,000 chips with the same configuration are necessary to explore the configuration file of an Actel A54SX16 chip with 24,000 system gates [Dipert, 2000]. Another aggravation of the attack is that only about 2–5% of all possible connections in an average design are actually used. In [Richard, 1998] a practical attack against AF FPGAs was performed and it was possible to alter one cell in two months at a cost of $1000. Based on these arguments some experts argue that physical attacks against AF FPGAs are harder to perform than against ASICs [Actel Corporation, 2002]. On the other hand, we know that AF FPGAs can be easily attacked if not connected to a power source. Hence, it is easier to drill holes to disconnect two connections or to repair destroyed layers. Also, depending on the source, the estimated cost of an attack and its complexity are lower [Richard, 1998].

Flash FPGAs. The connections in flash FPGAs are realized through flash transistors. That means the amount of electrons flowing through the gate changes after configuration and there are no optical differences as in the case of AF FP-GAs. Thus, physical attacks performed via analysis of the FPGA cell material are not possible. However, flash FPGAs can be analyzed by placing the chip in a vacuum chamber and powering it up. The attacker can then use a secondary electron microscope to detect and display emissions. The attacker has to get access to the silicon die, by removing the packet, before he can start the attack [Dipert, 2000]. However, experts are not certain about the complexity of such an attack and there is some controversy regarding its practicality [Actel Corporation, 2002, Richard, 1998]. Other possible attacks against flash FPGAs can be found in the related area of flash memory. The number of write/erase cycles are limited to 10,000–100,000, because of the accumulation of electrons in the floating gate causing a gradual rise of the transistors threshold voltage. This fact increases the programming time and eventually disables the erasing of the cell [Gutmann, 2001]. Another less common failure is the programming disturbance in which unselected erased cells gain charge when adjacent selected cells are written [Aritome et al., 1993]. This failure does not change the read operations but it can be detected with special techniques described in [Gutmann, 2001]. Furthermore, there are long term retention issues, like electron emission. The electrons in the floating gate migrate to the interface with the underlying oxide from where they tunnel into the substrate. This emission causes a net charge loss. The opposite occurs with erased cells where electrons are injected [Papadas et al., 1991]. Ionic contamination takes place as well but the influence on the physical behavior is so small that it can not be measured. In addition, hot carrier effects have a high influence, by building a tunnel between the bands. This causes a change in the threshold voltage of erased cells and is especially significant for virgin cells

[Haddad et al., 1989]. Another phenomenon is over-erasing, where an erase cycle is applied to an already-erased cell leaving the floating gate positively charged. Thus, turning the memory transistor into a depletion-mode transistor [Gutmann, 2001].

All the described effects change in a more or less extensive way the cell threshold voltage, gate voltage, or the characteristic of the cell. We remark that the stated phenomenons apply for EEPROM memory and that due to the structure of the FPGA cell these attacks can be simply adapted to attack flash/EEPROM FPGAs.

Summary of Physical Attacks. It is our position that due to the lack of published physical attacks against FPGAs, it is very hard (if at all possible) to predict the costs of such an attack. It is even more difficult to compare the effort needed for such an attack to a similar attack against an ASIC as there is no publicly available contribution which describes a physical attack against an FPGA that was completely carried out. Nevertheless, it is possible to draw some conclusions from our discussion above.

First, we notice that given the current size of state-of-the-art FPGAs, it seems infeasible, except perhaps for large government organizations and intelligence agencies, to capture the whole bitstream of an FPGA. Having said that, we should caution that in some cases, an attacker might not need to recover the whole bitstream information but rather a tiny part of it, e.g., the secret-key. This is enough to break the system from a practical point of view and it might be feasible.

On the other hand, there are certain properties that might increase the effort required for a physical attack against FPGAs when compared to ASICs. In the case of SRAM-, Flash-, EPROM-, and EEPROM-FPGAs there is no printed circuit (as in the case of ASICs) and therefore it is potentially harder to find the configuration of the FPGA. The attacker has to look for characteristics that were changed on the physical level during the programming phase. In the case of antifuse FPGAs the effort for a physical attack might increase compared to ASICs because one has to find a tiny connection with a diameter of about 100 nm in a 1 μm^2 insulation layer. Furthermore, only 2–5% of all possible connections are used in an average configuration.

21.3   Prevention of Attacks

This section shortly summarizes possible countermeasures that can be provided to minimize the effects of the attacks mentioned in the previous section. Most of them have to be realized by design changes by the FPGA manufacturers, but some could be applied during the programming phase of the FPGA.

21.3.1   How to Prevent Black Box Attacks

The Black Box Attack is not a real threat nowadays, due to the complexity of the designs and the size of state-of-the-art FPGAs (see Section 21.2.2). Furthermore, the nature of cryptographic algorithms prevents the attack as well. Cryptographic algorithms can be segmented in two groups: symmetric-key and public-key algorithms. Symmetric-key algorithms can be further divided into stream and block ciphers. Today's stream ciphers output a bit stream, with a period length of 128 bits [Thomas et al., 2003]. Block ciphers, like AES, are designed with a block length of 128 bits and a minimum key length of 128 bits. Minimum length in the case of public-key algorithms is 160 bits for ECC and 1024 bits for discrete logarithm and RSA-based systems. It is widely believed, that it is infeasible to perform a brute force attack and search a space with 2^{80} possibilities. Hence, implementations of these algorithms cannot be attacked with the black box approach.

21.3.2   How to Prevent Cloning of SRAM FPGAs

There are many suggestions to prevent the cloning of SRAM FPGAs, mainly motivated by the desire to prevent reverse engineering of general, i.e., non-cryptographic, FPGA designs. One solution would be to check the serial number before executing the design and delete the circuit if it is not correct. This approach is not practical because of the following reasons: 1) The whole chip, including the serial number can be easily copied; 2) Every board would need a different configuration; 3) Logistic complexity to manage the serial numbers [Kessner, 2000]. Another solution would be to use dongles to protect the design [Kean, 2001, Kessner, 2000]. Dongles are based on security-by-obscurity, and therefore do not provide solid security, as it can be seen from the software industry's experience using dongles for their tools. A more realistic solution would be to have nonvolatile memory and an FPGA in one chip or to combine both parts by covering them with epoxy. This reflects also the trend in chip manufacturing to have different components combined, e.g., the FPSLIC from Atmel. However, it has to be guaranteed that an attacker is not able to separate the parts.

Encryption of the configuration file is the most effective and practical countermeasure against the cloning of SRAM FPGAs. There are several patents that propose different scenarios related to the encryption of the configuration file: how to encrypt, how to load the file into the FPGA, how to provide key management, how to configure the encryption algorithms, and how to store the secret data In [Yip and Ng, 2000], the authors proposed that to partly decrypt the configuration file, in order to increase the debugging effort during the reverse engineering. If an attacker copies the partly decrypted file, the non-decrypted functionality is available, whereas the one decrypted is not. Thus, the attacker tries to find errors in the design unaware of the fact, that they are caused through the encrypted part of the configuration. Most likely an attacker with little resources, would have dropped the reverse engineering effort, when realizing that the parts are decrypted (which he did not do because he did not know). However, this approach adds hardly any extra complexity to an attack if we assume that an attacker has a lot of resources. In [Kelem and Burnham, 2000] an advanced scenario is introduced where the different parts of the configuration file are encrypted with different keys. The 60RS family from Actel was the first attempt to have a key stored in the FPGA in order to be able to encrypt the configuration file before transmitting it to the chip. The problem was that every FPGA had the same key on board. This implies that if an attacker has one key he can get the secret information from all FPGAs. In [Kean, 2001], the author discusses some scenarios where depending on the manufacturing cost, more than one key is stored in the FPGA.

An approach in a completely different direction would be to power the whole SRAM FPGA with a battery, which would make transmission of the configuration file after a power loss unnecessary. This solution does not appear practical, however, because of the power consumption of FPGAs. Hence, a combination of encryption and battery power provides a possible solution. Xilinx addresses this with an on-chip 3DES decryption engine in its Virtex II [Xilinx Inc., 2003] (see also [Pang et al., 2000]), where only the two keys are stored in the battery powered memory. Due to the fact that the battery powers only a very small memory cells, the battery is limited only by its own life span.

21.3.3   How to Prevent Readback Attacks

The readback attack can be prevented with the security bits set, as provided by the manufactures, see Section 21.2.4. If one wants to make sure that an attacker is not able to apply fault injection, the FPGA has to be embedded into a secure environment, where after detection of an interference the whole configuration is deleted or the FPGA is destroyed.

21.3.4   How to Prevent Side Channel Attack

In recent years, there has been a lot of work done to prevent side-channel attacks, see [Kaliski, Jr. et al., 2003]. The methods can generally be divided into software and hardware countermeasures, with the majority of proposals dealing with software countermeasures. "Software" countermeasures refer primarily to algorithmic changes, such as masking of secret keys with random values, which are also applicable to implementations in custom hardware or FPGA. Hardware countermeasures often deal either with some form of power trace smoothing or with transistor-level changes of the logic. Neither seem to be easily applicable to FPGAs without support from the manufacturers. However, some proposals such as duplicated architectures might work on today's FPGAs.

21.3.5 How to Prevent Physical Attacks

To prevent physical attacks, one has to make sure that the retention effects of the cells are as small as possible, so that an attacker cannot detect the status of the cells. Already after storing a value in a SRAM memory cell for 100–500 seconds, the access time and operation voltage will change [van der Pol and Koomen, 1990]. Furthermore, the recovery process is heavily dependent on the temperature: 1.5 hours at $75°C$, 3 days at $50°C$, 2 month at $20°C$, and 3 years at $0°C$ [Gutmann, 2001]. The solution would be to invert the data stored periodically or to move the data around in memory. Cryptographic applications also cause long-term retention effects in SRAM memory cells by repeatedly feeding data through the same circuit. One example is specialized hardware that always uses the same circuits to feed the secret key to the arithmetic unit [Gutmann, 2001]. Neutralization of this effect can be achieved by applying an opposite current [Tao et al., 1993] or by inserting dummy cycles into the circuit [Gutmann, 2001]. In terms of FPGA application, it is very costly or even impractical to provide solutions like inverting the bits or changing the location for the whole configuration file. A possibility could be that this is done only for the crucial part of the design, like the secret keys. Counter techniques such as dummy cycles and opposite current approach can be carried forward to FPGA applications.

In terms of flash/EEPROM memory cells, one has to consider that the first write/erase cycles causes a larger shift in the cell threshold [San et al., 1995] and that this effect will become less noticeable after ten write/erase cycles [Haddad et al., 1989]. Thus, one should program the FPGA about 100 times with random data, to avoid these effect (suggested for flash/EEPROM memory cells in [Gutmann, 2001]). The phenomenon of overerasing flash/EEPROM cells can be minimized by first programming all cells before deleting them.

21.4 Conclusions

This chapter analyzed possible attacks against the use of FPGAs in security applications. Black box attacks do not seem to be feasible for state-of-the-art FPGAs. However, it seems very likely for an attacker to get the secret information stored in a FPGA, when combining readback and fault injection attacks. Cloning of SRAM FPGAs and reverse engineering depend on the specifics of the system under attack and they will probably involve a lot of effort, but this does not seem entirely impossible. Physical attacks against FPGAs are very complex due to the physical properties of the semiconductors in the case of flash/SRAM/EEPROM FPGAs and the small size of AF cells. It appears that such attacks are even harder than analogous attacks against ASICs. Even though FPGA have different internal structures than ASICs with the same functionality, we believe that side-channel attacks against FPGAs, in particular power-analysis attacks, will be feasible too.

It seems from our previous remarks that, while the art of cryptographic algorithm implementation is reaching maturity, FPGAs as security platforms are not, and in fact, that they might be currently out of question for security applications. We do not think that is the right conclusion, however. It should be noted that many commercial ASICs with cryptographic functionality are also vulnerable to attacks similar to the ones discussed here. A commonly taken approach to prevent these attacks is to put the ASIC in a secure environment. A secure environment could, for instance, be a box with tamper sensors which triggers what is called "zeroization" of cryptographic keys, when an attack is being detected. Similar approaches are certainly possible for FPGAs too. (Another solution often taken by industry is not to care and to build cryptographic products with poor physical security, but we are not inclined to recommend this.)

References

Actel Corporation (2002). Design Security in Nonvolatile Flash and Antifuse. Avaialble at http://www.actel.com/appnotes/DesignSecurity.pdf.

Ajluni, C. (1995). Two New Imaging Techniques to Improve IC Defect Indentification. *Electronic Design*, 43(14):37–38.

Anderson, R. and Kuhn, M. (1997). Low Cost Attacks on Tamper Resistant Devices. In *5th International Workshop on Security Protocols*, pages 125–136. Springer-Verlag. LNCS 1361.

Aplan, J. M., Eaton, D. D., and Chan, A. K. (1999). Security Antifuse that Prevents Readout of some but not other Information from a Programmed Field Programmable Gate Array. United States Patent, Patent No. 5898776.

Aritome, S., Shirota, R., Hemink, G., Endoh, T., and Masuoka, F. (1993). Reliability Issues of Flash Memory Cells. *Proceedings of the IEEE*, 81(5):776–788.

Biham, E. and Shamir, A. (1997). Differential Fault Analysis of Secret Key Cryptosystems. In *Advances in Cryptology—CRYPTO '97*, pages 513–525. Springer-Verlag. LNCS 1294.

Boneh, D., DeMillo, R. A., and Lipton, R. J. (1997). On the Importance of Checking Cryptographic Protocols for Faults (Extended Abstract). In *Advances in Cryptology—EUROCRYPT '97*, pages 37–51. Springer-Verlag. LNCS 1233.

Dipert, B. (2000). Cunning circuits confound crooks. http://www.e-insite.net/ednmag/contents/images/21df2.pdf.

Guccione, S. A. and Levi, D. (2003). Jbits: A java-based interface to fpga hardware. Technical report, Xilinx Corporation, San Jose, CA, USA. Available at http://www.io.com/ guccione/Papers/Papers.html.

Gutmann, P. (1996). Secure Deletion of Data from Magnetic and Solid-State Memory. In *Sixth USENIX Security Symposium*, pages 77–90.

Gutmann, P. (2001). Data Remanence in Semiconductor Devices. In *10th USENIX Security Symposium*, pages 39–54.

Haddad, S., Chang, C., Swaminathan, B., and Lien, J. (1989). Degradations due to hole trapping in flash memory cells. *IEEE Electron Device Letters*, 10(3):117–119.

Kaliski, Jr., B. S., Koç, Ç. K., Naccache, D., Paar, C., and Walter, C. D., editors (2003). *CHES 1999-2003*, Berlin, Germany. Springer-Verlag. LNCS 1717/1965/2162/2523/2779.

Kean, T. (2001). Secure Configuration of Field Programmable Gate Arrays. In *International Conference on Field-Programmable Logic and Applications 2001 (FPL 2001)*, pages 142–151. Springer-Verlag. LNCS 2147.

Kelem, S. H. and Burnham, J. L. (2000). System and Method for PLD Bitstram Encryption. United States Patent, Patent Number 6118868.

Kessner, D. (2000). Copy Protection for SRAM based FPGA Designs. Available at http://www.free-ip.com/copyprotection.html.

Kocher, P., Jaffe, J., and Jun, B. (1999). Differential Power Analysis. In *CRYPTO '99*, pages 388–397. Springer-Verlag. LNCS 1666.

Kommerling, O. and Kuhn, M. (1999). Design Principles for Tamper-Resistant Smartcard Processors. In *USENIX Workshop on Smartcard Technology (Smartcard '99)*, pages 9–20.

Örs, S., Oswald, E., and Preneel, B. (2003). Power-Analysis Attacks on an FPGA — First Experimental Results. In *CHES 2003*, pages 35–50. Springer-Verlag. LNCS 2779.

Pang, R. C., Wong, J., Frake, S. O., Sowards, J. W., Kondapalli, V. M., Goetting, F. E., Trimberger, S. M., and Rao, K. K. (2000). Nonvolatile/ battery-backed key in PLD. United States Patent, Patent Number 6366117.

Papadas, C., Ghibaudo, G., Pananakakis, G., Riva, C., Ghezzi, P., Gounelle, C., and Mortini, P. (1991). Retention characteristics of single-poly EEPROM cells. In *European Symposium on Reliability of Electron Devices, Failure Physics and Analysis*, page 517.

Quisquater, J.-J. and Samyde, D. (2001). Electro Magnetic Analysis (EMA): Measures and Countermeasures for Smart Cards. In *International Conference on Research in Smart Cards, E-smart 2001*, pages 200 – 210, Cannes, France.

Richard, G. (1998). Digital Signature Technology Aids IP Protection. In EETimes—News. Available at http://www.eetimes.com/news/98/ 1000news/ digital.html.

San, K., Kaya, C., and Ma, T. (1995). Effects of erase source bias on Flash EPROM device reliability. *IEEE Transactions on Electron Devices*, 42(1):150–159.

Schroder, D. (1998). *Semiconducor Material and Device Characterization*. John Wiley and Sons, 2nd edition.

Seamann, G. (2000). FPGA Bitstreams and Open Designs. Available at http://www.opencollector.org/news/Bitstream.

Shang, L., Kaviani, A., and Bathala, K. (2002). Dynamic Power Consumption on the Virtex-II FPGA Family. In *2002 ACM/SIGDA 10th International Symposium on Field Programmable Gate Arrays*, pages 157–164. ACM Press.

Skorobogatov, S. and Anderson, R. (2002). Optical Fault Induction Attacks. In *CHES 2002*, pages 2–12. Springer-Verlag. LNCS 2523.

Soden, J. and Anderson, R. (1993). IC failure analysis: techniques and tools for quality and reliability improvement. *Proceedings of the IEEE*, 81(5):703–715.

Standaert, F.-X., van Oldeneel tot Oldenzeel, L., Samyde, D., and Quisquater, J.-J. (2003). Power Analysis of FPGAs: How Practical is the Attack. In *13th International Conference on Field Programmable Logic and Applications — FPL 2003*. Springer-Verlag. LNCS 2778.

Tao, J., Cheung, N., and Ho, C. (1993). Metal Electromigration Damage Healing Under Bidirectional Current Stress. *IEEE Transactions on Elecron Devices*, 14(12):554–556.

Thomas, S., Anthony, D., Berson, T., and Gong, G. (2003). The W7 Stream Cipher Algorithm. Available at http://www.watersprings.org/pub/id/draft-thomas-w7cipher-03.txt. Internet Draft.

van der Pol, J. and Koomen, J. (1990). Relation between the hot carrier lifetime of transistors and CMOS SRAM products. In *International Reliability Physics Symposium (IRPS 1990)*, page 178.

Williams, T., Kapur, R., Mercer, M., Dennard, R., and Maly, W. (1996). IDDQ Testing for High Performance CMOS—The Next Ten Years. In *IEEE European Design and Test Conference (ED&TC'96)*, pages 578–583.

Xilinx Inc. (2003). Using Bitstream Encryption. Handbook of the Virtex II Platform. Available at http://www.xilinx.com.

Yip, K.-W. and Ng, T.-S. (2000). Partial-Encryption Technique for Intellectual Property Protection of FPGA-based Products. *IEEE Transactions on Consumer Electronics*, 46(1):183–190.

Chapter 22

Bioinspired Stimulus Encoder for Cortical Visual Neuroprostheses

Leonel Sousa, Pedro Tomás[1], Francisco Pelayo[2], Antonio Martinez[2],
Christian A. Morillas[2] and Samuel Romero[2]

[1] *Department of Electrical and Computer Engineering, IST/INESC-ID, Portugal*
las@inesc-id.pt, pfzt@sips.inesc-id.pt

[2] *Department of Computer Architecture and Technology, University of Granada, Spain*
fpelayo@ugr.es, {amartinez, cmorillas, sromero}@atc.ugr.es

Abstract   This work proposes a real-time bioinspired visual encoding system for multielectrodes stimulation of the visual cortex supported on Field Programmable Logic. This system includes the spatio-temporal preprocessing stage and the generation of time-varying spike patterns to stimulate an array of microelectrodes and can be applied to build a portable visual neuroprosthesis. It only requires a small amount of hardware thanks to the high operating frequency of modern FPGAs, which allows to sequentialize some of the required processing. Experimental results show that, with the proposed architecture, a real-time visual encoding system can be implemented in FPGAs with modest capacity.

Keywords: Visual cortical-neuron stimulation, retina model, FPL implementation, visual information processing

22.1   Introduction

Nowadays, the design and the development of visual neuroprostheses interfaced with the visual cortex is being tried to provide a limited but useful visual sense to profoundly blind people. The work presented has been carried out within the EC project "Cortical Visual Neuroprosthesis for the Blind" (CORTIVIS), which is one of the research initiatives for developing a visual neuroprosthesis for the blind [1, 2].

P. Lysaght and W. Rosenstiel (eds.),
New Algorithms, Architectures and Applications for Reconfigurable Computing, 279–290.
© 2005 *Springer. Printed in the Netherlands.*

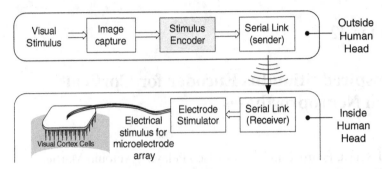

Figure 22.1. Cortical visual neuroprosthesis.

A block diagram of the cortical visual neuroprosthesis is presented in Fig. 22.1. It includes a programmable visual stimulus encoder, which computes a predefined retina model, to process the input visual stimulus and to produce output patterns that approximate the spatial and temporal spike distributions required for effective cortical stimulation. These output patterns are represented by pulses that are mapped on the primary visual cortex and are coded by using Address Event Representation [3]. The corresponding signals are modulated and sent through a Radio Frequency (RF) link, which also carries power, to the electrode stimulator. This project uses the Utah microelectrode array [4], which consists on an array of 10×10 silicon microelectrodes separated by about 400 μm in each orthogonal direction (arrays of 25×25 microelectrodes are also considered). From experimental measures on biological systems, it can be established a time of 1 ms to "refresh" all the spiking neurons, which means an average time slot of 10 μs dedicated to each microelectrode. The RF link bandwidth allows communication at a bit-rate of about 1 Mbps, which means an average value of 10 kbps for each microelectrode in a small size array of 10×10 electrodes or about 1.6 kbps for the 25×25 microelectrode array.

The work presented herein addresses the design of digital processors for implementing the block shown shaded in Fig. 22.1 and extended in Figure 22.2 in Field Programmable Logic (FPL) technology. The adopted retina model is a complete approximation of the spatio-temporal receptive fields characteristic

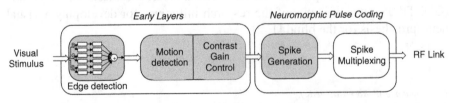

Figure 22.2. Retina early layers.

response of the retina ganglion cells. It also includes two other blocks. A leaky integrate-and-fire block, that generates the spikes to stimulate the cortical cells, i.e. the neuromorphic pulse coding. A spike channel interface block which uses the Address Event Representation (AER) protocol to communicate information about the characteristics of spikes and addresses of target microelectrodes without timestamps. The architecture of the system has been designed taking into account the specifications of the problem described above and the technical characteristics of modern Field Programmable Gate Array (FPGA) devices. Experimental results show that a complete artificial model that generates neuromorphic pulse-coded signals can be implemented in real-time even in FPGAs with low-capacity.

The rest of this paper is organized as follows. The architecture for modelling the retina and for coding the event lists that will be carried out to the visual cortex is presented in section 22.2. Section 22.3 describes the computational architectures designed for implementing the retina model in FPL, discussing their characteristics and suitability to the technology. Section 22.4 presents experimental results obtained by implementing the stimulus encoder on a FPGA and section 22.5 concludes the presentation.

22.2   Model Architecture

The neuron layers of the human retina perform a set of different tasks, which culminate in the spiking of ganglion cells at the output layer [5]. These cells have different transient responses, receptive fields and chromatic sensibilities. The system developed implements the full model of the retina, which includes all the processing layers, plus the protocol for carrying the spikes to the visual cortex in a serial way through a RF link.

The visual stimulus encoder includes two bio-inspired blocks which are shown in grey in Fig. 22.2. The first is an edge detection block based on a set of weighted Gaussian spatial filters dependent on time through low pass filters. The second block consists of a local contrast gain controller (CGC) which attempts to compensate for the biological delay of the visual signal processing and transmission. The spike generation block produces the spike events which will stimulate the visual cortex.

22.2.1   Retina Early Layers

The presented retina computational model is based on research published in [5, 6, 7, 8] and was extended to the chromatic domain by considering independent filters for the basic colors.

The first filtering element, the retina early layers (see Fig. 22.2) is an edge detector composed of a set of two Gaussian spatial filters per color channel, parameterized by a gain a_{ij} and a standard deviation σ_{ij}, as shown in equation 22.1.

$$g_i(\mathbf{r}) = \frac{a_i}{\sqrt{2\pi}\sigma_i} e^{-\frac{r^2}{2\sigma_i^2}} \tag{22.1}$$

In order to detect edges the Gaussian filters of each color channel are parameterized with opposite signs and different standard deviations. The output is then processed by a set of temporal first-order low pass filters:

$$h_{ij}(t) = H(t) \cdot B_{ij} e^{-B_{ij} \cdot t} \tag{22.2}$$

where $H(t)$ represents the Heaviside step function and B_{ij} is the decay rate. The output of the edge detection block is:

$$m(\mathbf{r}, t) = \sum_{i=R,G,B} s_i(\mathbf{r}, t) * \sum_{j=1}^{2} g_{ij}(\mathbf{r}) \cdot h_{ij}(t) \tag{22.3}$$

where $*$ denotes the convolution operator and $\{s_R, s_G, s_B\}$ are the color components of the visual stimuli. The bilinear approximation was applied to derive a digital version of the time filters represented in the Laplace domain [9]. It leads to first order Infinite Impulse Response (IIR) digital filters with input $v_{ij}[\mathbf{q}, t]$ and output $w_{ij}[\mathbf{q}, t]$ which can be represented by equation 22.4.

$$w_{ij}[\mathbf{q}, n] = b_{LPij} \times w_{ij}[\mathbf{q}, n-1] + c_{LPij} \times (v_{ij}[\mathbf{q}, n] + v_{ij}[\mathbf{q}, n-1]) \tag{22.4}$$

The second filtering block of the visual stimulus encoder is a motion detector based on a temporal high pass filter defined by a pole in $-\alpha$ whose impulse response is represented in equation 22.5.

$$h_{HP}(t) = \delta(t) - \alpha H(t) e^{-\alpha t} \tag{22.5}$$

The filter was also represented in the digital domain by applying the bilinear approximation. The result is a first order Infinite Impulse Response (IIR) digital filter which can be represented by equation 22.6 where $m[\mathbf{q}, n]$ is the filter input and $u[\mathbf{q}, n]$ is the output.

$$u[\mathbf{q}, n] = b_{HP} \times u[\mathbf{q}, n-1] + c_{HP} \times (m[\mathbf{q}, n] - m[\mathbf{q}, n-1]) \tag{22.6}$$

The resulting activation $u(\mathbf{r},t)$ is multiplied by a Contrast Gain Control (CGC) modulation factor $g(\mathbf{r}, t)$ and rectified to yield the firing rate of the ganglion cells response to the input stimuli.

The CGC models the strong modulatory effect exerted by stimulus contrast. The CGC non-linear approach is also used in order to model the "motion anticipation" effect observed on experiments with a continuous moving bar [7]. The CGC feedback loop involves a low-pass temporal filter with the following impulse response:

$$h_{LP}(t) = Be^{-\frac{t}{\tau}} \tag{22.7}$$

where B and τ define the strength and constant time of the CGC. The filter is computed in the digital domain, by applying the same approximation as for the high-pass filter, by equation 22.8.

$$z[\mathbf{q}, n] = b_{LP} \times z[\mathbf{q}, n-1] + c_{HP} \times (y[\mathbf{q}, n] + y[\mathbf{q}, n-1]) \tag{22.8}$$

and the output of the filter is transformed into a local modulation factor (g) via the non-linear function:

$$k(t) = \frac{1}{1 + [v(t) \cdot H(v(t))]^4} \tag{22.9}$$

The output is rectified by using the function expressed in eq. 22.10:

$$F_i(\mathbf{r}, t) = \psi H(y(\mathbf{r}, t) + \theta)[y(\mathbf{r}, t) + \theta] \tag{22.10}$$

where ψ and θ define the scale and baseline value of the firing rate $f_i(\mathbf{r}, t)$.

The system to be developed is fully programmable and typical values for all parameters of the model are found in [6]. The model has been completely simulated in MATLAB, by using the *Retiner* environment for testing retina models [10].

22.2.2   Neuromorphic Pulse Coding

The neuromorphic pulse coding block converts the continuous-varying time representation of the signals produced in the early layers of the retina into a neural pulse representation. In this new representation the signal provides new information only when a new pulse begins. The model adopted is a simplified version of an integrate-and-fire spiking neuron [11]. As represented in Fig. 22.3, the neuron accumulates input values from the respective receptive field (output firing rate determined by retina early layers) until it reaches a given threshold. Then it fires and discharges the accumulated value. A leakage term is included to force the accumulated value to diminish for low or null input values. The

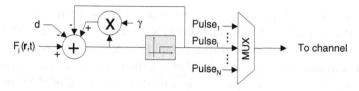

Figure 22.3. Neuromorphic pulse coding.

pulses are then generated accordingly to equations 22.11 and 22.12

$$P_{acc}[\mathbf{q}, n] = F[\mathbf{q}, n] + \gamma \cdot P_{acc}[\mathbf{q}, n-1] - pulse[\mathbf{q}, n-1] - d \quad (22.11)$$

$$pulse[\mathbf{q}, n] = H(P_{acc}[\mathbf{q}, n] - \phi) \quad (22.12)$$

where P_{acc} is the accumulated value, F is the early layer's output, $\gamma < 1$ sets the decay rate of P_{acc} ($decay = 1 - \gamma$) and H represents the Heaviside step function. The circuit fires a pulse whenever the accumulated value P_{acc} is bigger then the threshold ϕ.

The amplitude and duration of the pulses are then coded using AER, which is represented in a simplified way in Fig. 22.3 by a multiplexer. This information is then sent to the corresponding microelectrodes via the RF link. An event consists of an asynchronous bus transaction that carries the address of the corresponding microelectrode and is sent at the moment of pulse onset (no timestamp information is communicated). An arbitration circuit is required at the sender side and the receiver has to be listening to the data link with a constant latency.

22.3 FPL Implementation

This section discusses the implementation in FPL technology of the visual encoding system presented in the previous section. The usage of FPL technology will allow changing the model parameters without the need of designing another circuit. This has special importance since different patients have different sight parameters therefore requiring adjustments to the artificial retina. FPL may also allow changing model blocks if new information on how visual stimuli is processed reveals the need to introduce new filtering blocks.

For the design of the system, different computational architectures were considered both for the retina early layers and for the neuromorphic coding of the pulses. These architectures lead to implementations with different characteristics, in terms of hardware requirements and speed. The scalability and programmability of the system are also relevant aspects that are taken into account for FPL implementations.

22.3.1 The Retina Early Layers

There can be multiple approaches to implementing the retina early layers each of which involve different hardware requirements, so the first step would be to analyze the necessary hardware to build the desired processor. Assuming a typical convolutional kernel of 7×7 elements for the Gaussian spatial filters and that high-pass and low-pass temporal filters are computed by using the difference equations 22.4, 22.6 and 22.8, respectively, then 104 multiplications and 107 additions per image cell are required for just one spatial channel. For a matrix of 100 microelectrodes, the hardware required by a fully parallel architecture makes it impractical in FPL technology. Therefore, the usage of the hardware has to be multiplexed in time, but the temporal restriction of processing the whole matrix with a frequency up to 100 Hz must also be fulfilled. This restriction is however wide enough to consider the processing of cells with a full multiplexing schema inside each main block of the architecture model presented in section 12.2: edge detection and motion detection and anticipation.

The architectures of these blocks can then be described as shown in Figure 22.4, where each main block includes one or more RAM blocks to save processed frames in intermediate points of the system in order to compute the recursive time filters. At the input, dual-port RAMs are used for the three color channels, while a ROM is used to store the coefficient tables for the Gaussian spatial filters. The operation of the processing modules is locally controlled and the overall operation of the system is performed by integrating the local control in a global control circuit. The control circuit is synthesized as a Moore state machine, because there is no need to modify the output asynchronously with variations of the inputs. All RAM address signals, not shown in the Figure 22.4, are generated by the control circuit.

The Gaussian filter module (see Fig. 22.4a) calculates the 2D convolution between the input image and the predefined matrix coefficients stored in the ROM. It operates in a sequential method, where pixels are processed one at a time in a row major order and an accumulator is used to store the intermediate and final results of the convolution. The local control circuit stores the initial

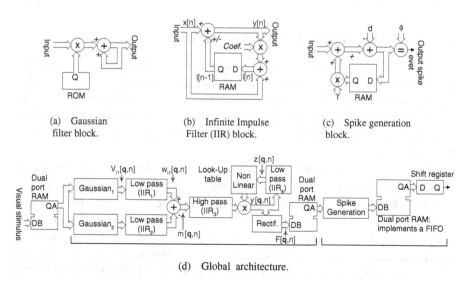

(a) Gaussian filter block.

(b) Infinite Impulse Filter (IIR) block.

(c) Spike generation block.

(d) Global architecture.

Figure 22.4. Detailed diagram for the visual encoder computational architecture.

pixel address, row and column, and then successively addresses all the data, and respective filtered coefficients stored in the ROM, required for the computation of the convolution. After multiplying each filter coefficient the resulting value is successively accumulated. When the calculation is finished for a given pixel (see Fig. 22.4d), the value is sent to the low pass filter. The filter (see Fig. 22.4b) is computed in a transposed form where the output is calculated by adding the input $x[n]$ to the stored value $l[n]$ which was previously computed as

$$l[n-1] = b_{LPij} \cdot y[n-1] + x[n-1] \qquad (22.13)$$

The input scaling by C_{LPij} as shown in equation 22.4 is made with the Gaussian filter coefficient matrix. The output of the low pass filters of the different color channels is then added accordingly to eq. 22.3 and the result, $m[q, n]$, is processed by a temporal high pass filter module also implemented with the IIR module of Fig. 22.4 before being processed by the CGC block. This module consists on a low pass filter (eq. 22.8) and a non-linear function (eq. 22.9) which is computed by using a lookup table stored in a RAM block. The low-pass filter circuit is again implemented with the IIR module of Fig. 22.4(b). The last processing module is a rectifier which clips the signal whenever its value decreases below a predefined threshold. It is implemented by a simple binary comparator whose output is used to control a multiplexer.

The overall circuit operates in a 4 stage pipeline corresponding to each one of the main processing blocks. Only the first pipeline stage requires a variable number of clock cycles depending on the dimension of the filter kernel–49 cycles for a 7×7 kernel. Each of the two other pipeline stages is solved in a single clock cycle.

22.3.2   Neuromorphic Pulse Coding

Fig. 22.4d shows two processing modules associated with the neuromorphic pulse coding: *i)* for pulse generation and its representation and *ii)* to arbitrate the access to the serial bus (the input to the serial link in Figure 22.1). This block is connected to the retina early layers through a dual port RAM (CGC block in Fig. 22.4) where one writes data onto one port and the other reads it from the second port.

The pulse generation circuit, which converts from firing rate to pulses, can be seen as a simple Digital Voltage Controlled Oscillator (DVCO) working in a two clock cycle stage pipeline. In the first stage the input firing rate is added to the accumulated value. In the second stage a leakage value is subtracted and, if the result is bigger than the threshold ϕ, a pulse is fired and the accumulator returns to the zero value (see Fig. 22.4c). Note that in this architecture the accumulator circuit is made with a RAM, since it corresponds to a single adder and multiple accumulator registers for the different microelectrodes.

AER was used to represent the pulses while the information is serialized. In a first approach, the architecture consisted of an arbitration tree to multiplex the onset of events (spikes) onto a single bus. In this tree, we check if one or more inputs has a pulse to be sent and arbitrate the access to the bus through the request/acknowledge signals in the tree [3]. This tree consists of multiple subtrees, where registers can be used for buffering the signals between them. This architecture is not scalable, since it requires a great amount of hardware and is not a solution when arrays with a great number of microelectrodes are used (see section 12.4). To overcome this problem, and since the circuit for pulse generation is intrinsically sequential, a First In First Out (FIFO) memory is used to register the generated spikes for the microelectrodes. Only one pulse is generated in a clock cycle and information about it is stored in the FIFO memory because requests may find the communication link busy. This implementation has the advantage of not increasing the hardware resources required with the increase in the number of microelectrodes, but its performance is more dependent on the channel characteristics since it increases the maximum wait time to send a pulse.

The FIFO is implemented by using a dual-port data RAM, where input is stored in one port and, at the other port, the pulse request is sent to the channel. Also to avoid overwriting when reading data from the FIFO a shift register is used to interface the data output of the RAM with the communication channel.

22.4   Experimental Results

The visual encoding system was described in VHDL and exhaustively tested on a Digilent DIGILAB II board based on a XILINX SPARTAN XC2S200 FPGA [12] with modest capacity. The synthesis tool used was the one supplied by the FPGA manufacturer, the ISE WebPACK 5. This FPGA has modest capacities, with 56 kbit of block-RAM and 2352 slices, corresponding to a total of 200,000 system-gates. The functional validation of the circuits was carried out by comparing the results obtained with MATLAB Retiner [10].

For the test and experimental results presented in this section, only one color channel is considered and the input signal is represented in 8-bit grayscale. However, internally 12-bit signals are used in order to reduce discretization errors. The dimension of the microelectrode array is 100.

Analyzing the results in Table 22.1 it is clearly seen that the retina early layers do not require many FPGA slices but consumes a significant amount of memory. The synthesis of the circuit for an array of 1024 microelectrodes shows a similar percentage of FPGA slice occupation but it uses all the 14 RAM blocks supplied by the SPARTAN XC2S200. In terms of time restrictions, the solution works very well since it can process the array of input pixels in a short time, about 0.1ms for 100 microelectrodes, which is much lower than the specified maximum of 10ms. In fact this architecture allows to process movies at frame rate of 100 Hz with 10,000 pixels per image without increasing the number of slices used.

The analysis of the AER translation method has however different aspects. For a typical matrix of 100 microelectrodes, the tree solution (with or without intermediate registers) is a valid one as the resource occupation is low. However, the increase in the number of microelectrodes implies the usage of a great amount of hardware and this solution becomes impracticable for FPL technology. In that case, the proposed FIFO based solution has the great advantage of accommodating a great number of electrodes with a small amount of hardware and just one RAM block. With a channel bandwidth of about 1 Mbps, the FIFO based AER circuit does not introduce any temporal restrictions in the firing rate. Even when operating at a lower clock frequency than that admitted by the retina early layers circuit, e.g. 40 MHz, the FIFO based circuit is able to process 100 requests in about 2.5 μs which is much less than the value of 1 ms initially specified and also much less than the time required to send the information through the channel.

In Table 22.1 results are individually presented for the retina early layers and for the neuromorphic pulse coding circuits. The overall system was also synthesized and implemented in a SPARTAN XC2S200 FPGA. In this case the Neuromorphic Pulse Coding RAM block had to be implemented in distributed RAM due to the low memory capacity of the FPGA. A clock frequency greater than 40 MHz is achieved by using about 28% of the slices and 12 of the total of 14 RAM-blocks, for 100 microelectrodes.

Table 22.1. Retina encoding system implemented on a SPARTAN XC2S200 FPGA.

Block	Number of microelectrodes	Slice occupation	Maximum clock frequency	Number of RAM blocks used
Retina early layers	100	21%	47 MHz (49 cc*)	5
	1024	21%	47 MHz (49 cc*)	10
Spike Generation	100	2%	51 MHz	1
	1024	2%	51 MHz	3**
AER	100	9%	99 MHz	0
Registered Tree	256	17%	98 MHz	0
	512	37%	93 MHz	0
AER	100	10%	64 MHz	0
Unregistered Tree	256	21%	63 MHz	0
	512	45%	57 MHz	0
FIFO AER	***	5%	75 MHz	1

* 49 clock cycles are required for processing a pixel
** implemented as distributed RAM in the overall system occupying 33% extra slices
*** not dependent of the number of micro-electrodes

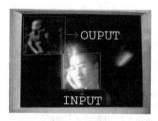

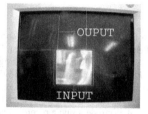

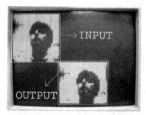

(a)   Edge detecting.   (b)   Early Layers.   (c)   Spike generation.

Figure 22.5.   Testing of the visual encoding module for a microelectrode array of 32 × 32.

In order to test the visual encoding module a system prototype was developed based on two XILINX XC2S200 FPGAs (one to implement the visual encoder and the other to implement the capture of image frames and visualization of the system output), one digital output color camera module, model C3188A [13] which captures images in QVGA at a frame rate up to 60Hz and one standard monitor. Figure 22.5 shows some of the experimental results of the Visual Encoding implementation: Fig. 22.5(a) show the result of the edge detecting block, Fig. 22.5(b) shows the output of the Early Layers module, where moving fingers caused a non-zero output. Fig. 22.5(c) shows the output of the neuromorphic pulse coding module after image restoration with a second order low pass filter with poles at 6 Hz.

22.5   Conclusions

The present work proposes a computational architecture for implementing a complete model of an artificial retina in FPL technology. The system is devised to stimulate a number of intra-cortical implanted microelectrodes for a visual neuroprosthesis. It performs a bio-inspired processing and encoding of the input visual information. The proposed processing architecture is designed to fulfill the constraints imposed by the telemetry system to be used, and with the requirement of building a compact and portable hardware solution.

The synthesis results demonstrate how the adopted pipelined and time multiplexed approach makes the whole system fit well on a relatively low complexity FPL circuit, while real-time processing is achieved. The use of FPL is also very convenient to easily customize (re-configure) the visual pre-processing and encoding system for each implanted patient.

Acknowledgments

This work has been supported by the European Commission under the project CORTIVIS ("Cortical Visual Neuroprosthesis for the Blind", QLK6-CT-2001-00279).

References

[1] Cortical visual neuro-prosthesis for the blind (cortivis): http://cortivis.umh.es.

[2] Ahnelt P., Ammermüller J., Pelayo F., Bongard M., Palomar D., Piedade M., Ferrandez J., Borg-Graham L., and Fernandez E. Neuroscientific basis for the design and development of a bioinspired visual processing front-end. In *Proc. of IFMBE*, pages 1692–1693, Vienna, 2002.

[3] Lazzaro J. and Wawrzynek J. A multi-sender asynchronous extension to the address event protocol. In *Proc. of 16th Conference on Advanced Research in VLSI*, pages 158–169, 1995.

[4] Maynard E. The Utah intracortical electrode array: a recording structure for potential brain-computer interfaces. *Elec. Clin. Neurophysiol.*, 102:228–239, 1997.

[5] Wandell Brian. *Foundations of Vision: Behavior, Neuroscience and Computation*. Sinauer Associates, 1995.

[6] Wilke S., Thiel A., Eurich C., Greschner M., Bongard M., Ammermüler J., and Schwegler H. Population coding of motion patterns in the early visual system. *J. Comp Physiol A*, 187:549–558, 2001.

[7] Berry M., Brivanlou I., Jordan T., and Meister M. Anticipation of moving stimuli by the retina. *Nature (Lond)*, 398:334–338, 1999.

[8] Markus Meister and Michael J. Berry II. The neural code of the retina. *Neuron*, 22:435–450, March 1999.

[9] Oppenheim A. and Willsky A. *Signal and Systems*. Prentice Hall, 1983.

[10] F.J. Pelayo, A. Martìnez, C. Morillas, S. Romero, L. Sousa, and P. Tom·s. Retina-like processing and coding platform for cortical neuro-stimulation. In *Proc. of 25th Annual IEEE International Conference of the Engineering in Medicine and Biology Society*, IEEE Catalog number 03CH37439C, pages 2023–2026, Cancun, Mexico, September 2003.

[11] Gerstner W. and Kistler W. *Spiking Neuron Models*. Cambridge University Press, 2002.

[12] XILINX. *Spartan-II 2.5V FPGA Family: Functional Description*, 2001. Product Specification.

[13] Quasar Electronics Ltd. C3188A - Digital Output Colour Camera Module. http://www.electronic-kits-and-projects.com/kit-files/cameras/d-c3188a.pdf.

Chapter 23

A Smith-Waterman Systolic Cell

C.W. Yu, K.H. Kwong, K.H. Lee and P.H.W. Leong

Department of Computer Science and Engineering
The Chinese University of Hong Kong, Shatin, Hong Kong
y_chi_wai@hotmail.com,edwardkkh@alumni.cuhk.net,
khlee@cse.cuhk.edu.hk,phwl@cse.cuhk.edu.hk

Abstract   In order to understand the information encoded by DNA sequences, databases containing large amount of DNA sequence information are frequently compared and searched for matching or near-matching patterns. This kind of similarity calculation is known as sequence alignment. To date, the most popular algorithms for this operation are heuristic approaches such as BLAST and FASTA which give high speed but low sensitivity, i.e. significant matches may be missed by the searches. Another algorithm, the Smith-Waterman algorithm, is a more computationally expensive algorithm but achieves higher sensitivity. In this paper, an improved systolic processing element cell for implementing the Smith-Waterman on a Xilinx Virtex FPGA is presented.

Keywords: Field programmable gate arrays, Systolic arrays, Biochemistry.

23.1   Introduction

Bioinformatics is becoming an increasingly important field of research. With the ability to rapidly sequence DNA information, biologists have the tools to, among other things, study the structure and function of DNA; study evolutionary trends; and correlate DNA information with disease. For example, two genes were identified to be involved in the origins of breast cancer in 1994 [1]. Such research is only possible with the help of high speed sequence comparison.

All the cells of an organism consist of some kind of genetic information. They are carried by a chemical known as the deoxyribonucleic acid (DNA) in the nucleus of the cell. DNA is a very large molecule and nucleotide is the basic unit of this type of molecule. There are 4 kinds of nucleotides and each

P. Lysaght and W. Rosenstiel (eds.),
New Algorithms, Architectures and Applications for Reconfigurable Computing, 291–300.
© 2005 *Springer. Printed in the Netherlands.*

have different bases, namely adenine, cytosine, guanine and thymine. Their abbreviated forms are "A", "C", "G" and "T" respectively. In this paper, the sequence is referred to as a string, and the bases form the alphabet for the string.

It is possible to deduce the original sequencing in DNA which codes a particular amino acid. By finding the similarity between a number of "amino-acid producing" DNA sequences and a genuine DNA sequence of an individual, one can identify the protein encoded by the DNA sequence of the individual. In addition, if biologists succeed in finding the similarity between DNA sequences of two different species, they can understand the evolutionary trend between them. Another important usage is that the relation between disease and inheritance can also be studied. This is done by aligning specific DNA sequences of individuals with disease to those of normal people. If correlations can be found which can be used to identify those susceptible to certain diseases, new drugs may be made or better techniques invented to treat the disease. There are many other applications of bioinformatics and this field is expanding at an extremely fast rate.

A human genome contains approximately 3 billion DNA base pairs. In order to discover which amino acids are produced by each part of a DNA sequence, it is necessary to find the similarity between two sequences. This is done by finding the minimum string edit distance between the two sequences and the process is known as sequence alignment.

There are many algorithms for doing sequence alignment. The most commonly used ones are FASTA [2] and BLAST [3]. BLAST and FASTA are fast algorithms which prune the search involved in a sequence alignment using heuristic methods. The Smith-Waterman algorithm [4] is an optimal method for homology searches and sequence alignment in genetic databases and makes all pairwise comparisons between the two strings. It achieves high sensitivity as all the matched and near-matched pairs are detected, however, the computation time required strongly limits its use.

Sencel Bioinformatics [5] compared the sensitivity and selectivity of various searching methods. The sensitivity was measured by the coverage, which is the fraction of correctly identified homologues (true positives). The coverage indicates what fraction of structurally similar proteins one may expect to identify based on sequence alone. Their experiments show that for a coverage around 0.18, the errors per query of BLAST and FASTA are about two times that of the Smith-Waterman algorithm.

Many previous ASIC and FPGA implementations of the Smith-Waterman algorithm have been proposed and some are reviewed in Section 23.4. To date, the highest performance chip [6] and system level [7] figures have been achieved using a runtime reconfigurable implementation which directly writes one of the strings into the FPGA's bitstream.

In this work, an FPGA-based implementation of the Smith-Waterman algorithm is presented. The main contribution of this work is a new design for a Smith-Waterman cell that uses three Xilinx Virtex slices and is able to achieve the same density and performance as an earlier reported cell [6], without the need to perform runtime reconfiguration. This has advantages in that the design is less FPGA device specific and thus can be used for non-Xilinx FPGA devices as well as ASICs. Whereas the runtime reconfigurable design requires JBits, a Xilinx specific API for runtime reconfiguration, the design presented in this paper was written in standard VHDL. Moreover, in the proposed design, both strings being compared can be changed rapidly as compared to a runtime reconfigurable system in which the bitstream must be generated and downloaded, which is typically a very slow process since a large bitstream must be manipulated and downloaded via a slow interface. This reconfiguration process may become a bottleneck, particularly for small databases. Furthermore, other applications may require both strings to change quickly. The design was implemented and verified using Pilchard [8], a memory-slot based reconfigurable computing environment.

23.2 The Smith-Waterman Algorithm

The Smith-Waterman Algorithm is a dynamic programming technique which utilizes a 2D table. As an example of its application, suppose that one wishes to compare sequence S ("ACG") with sequence T ("ATC"). The intermediate values a, b and c (shown in Fig. 23.2) are then used to compute d according to the following formula:

$$d = min \begin{cases} \begin{cases} a & if\ S_i = T_j \\ a + sub & if\ S_i \neq T_j \end{cases} \\ b + ins \\ c + del \end{cases} \tag{23.1}$$

If the strings being compared are the same, the value a is used for d. Otherwise, the minimum of a plus some substitution penalty sub, b plus some insertion penalty ins and c plus some deletion penalty del is used for d. Data dependencies mean that entries d in the table can only be calculated if the corresponding a, b, c values are already known and so the computation of the table spreads out from the origin as illustrated in Fig. 23.1. As an example, the first entry that can be computed is that for "AA" in Fig. 23.2(a). Since $S_i = T - i =$ 'A', according to Equation. 23.1, $d = a$ and so the entry is set to 0. In order to complete the table, the template of Fig. 23.2(b) is moved around the table constrained by the dependencies indicated by Fig. 23.1.

The substitution, insertion and deletion penalties can be adjusted for different comparison requirements. If the presence of redundant characters is relatively

Table 23.1. Figure showing the progress of the Smith-Waterman algorithm, when the string "ACG" is compared with "ATC".

		A	C	G
	0	1	2	3
A	1	?	?	?
T	2	?	?	?
C	3	?	?	?

(a) Initial table.

		S_i	
		a	b
T_i		c	d

(b) Equation 23.1 values.

		A	C	G
	0	1	2	3
A	1	0	1	2
T	2	1	2	3
C	3	2	1	2

(c) Final table.

less acceptable than just a difference in characters, the insertion and deletion penalties can be set to a higher value than the substitution penalty. In the alignment system presented, the insertion and deletion penalties were fixed at 1 and the substitution penalty set to 2, as is typical in many applications.

If S and T are m and n in length respectively, then the time complexity of a serial implementation of the Smith-Waterman algorithm is $O(mn)$. After all the squares have been processed, the result of Fig. 23.2(c) is obtained. In a parallel implementation, the positive slope diagonal entries of Fig. 23.1 can be computed simultaneously. The final edit distance between the two strings appears in the bottom right table entry.

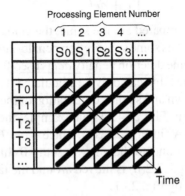

Figure 23.1. Data dependencies in the alignment table. Thick lines show entries which can be computed in parallel and the time axis is arranged diagonally.

23.3 FPGA Implementation

In 1985, Lipton and Lopresti observed that the values of b and c in Equation 23.1 are restricted to $a \pm 1$ and the equation can be simplified to obtain [9]:

$$d = \begin{cases} a & \text{if } ((b \text{ or } c) = a - 1) \text{ or } (S_i = T_j) \\ a+2 & \text{if } ((b \text{ and } c) = a + 1) \text{ and } (S_i \neq T_j) \end{cases} \tag{23.2}$$

Using Equation 23.2, it can be seen that the data elements b, c and d only have two possible values. Therefore, the number of data bits used for the representation of b, c and d can be reduced to 1 bit. Furthermore, two bits can be used to represent the four possible values of the alphabet.

The processing element (PE) shown in Fig. 23.2 was used to implement Equation 23.2. A number of PEs are then connected in a linear systolic array to process diagonal elements in the table in parallel. As shown in Fig. 23.1, PEs are arranged horizontally and are responsible for their corresponding column. In the description that follows, the sequence that changes infrequently is S and the sequences from the database are T. In each PE, two latches are used to store a character S_i. These characters are shifted and distributed to every processing element before the actual comparison process beings. The base pairs of T are passed through the array during the comparison process, during which the d of Equation 23.2 is also computed.

In order to calculate d, inputs a, b and c should be available. In the actual implementation, the new values b, c and d are calculated during the comparison of the characters as follows:

1 data_in is the new value of c and it is stored in a flip-flop. At the same time, this new c value and the previous d value (from a flip-flop) determines the new b value ($b = temp\_c$ XNOR $temp\_d$)

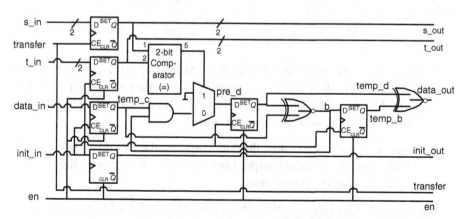

Figure 23.2. The Smith-Waterman processing element (PE). Boxes represent D-type flip-flops.

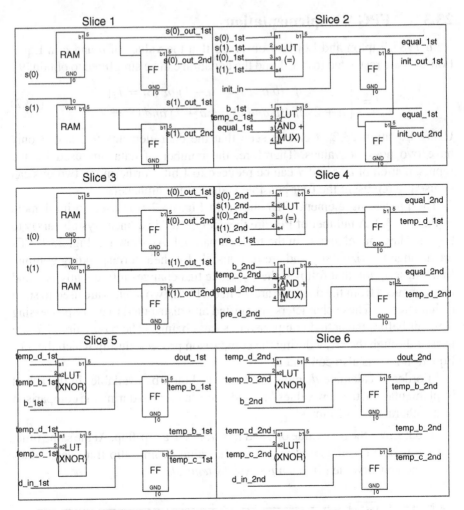

Figure 23.3. Two processing elements mapped to 6 Virtex slices.

2 The new *b* value is stored in a flip-flop. At the same time, the output of a 2-to-1 MUX is then selected depending on whether $S_i = T_i$. The output of the MUX (a '0' value or (*b* AND *temp_c*)) becomes the new *d* value. This new *d* value is stored in a flip-flop.

3 Values of *b* and *d* determine the data output of the PE (*data_out=temp_b* XNOR *temp_d*). The data output from this PE is connected to the next PE as its data input (its new *c* value)

When the *transfer* signal is high, the sequence *S* is shifted through the PEs. When the *en* signal is high, all the flip-flops (except the two which store the string *S*) are reset to their initial values. The *init* signal is high when new signals from the preceding PE are input and the new value of d calculated. When the *init* signal is low, the data in all the flip-flops are unchanged.

Each PE used 8 flip-flops as storage elements and 4 LUTs to implement the combinational logic. Thus the design occupied 4 Xilinx Virtex slices. Guccione and Keller [6] used runtime reconfiguration to write one of the sequences into the bitstream, saving 2 flip-flops and implementing a PE in 3 slices. In the proposed approach, two otherwise unused LUTs were configured as shift register elements using the Xilinx distributed RAM feature [10]. Thus the design occupies 3 Xilinx Virtex slices per PE, without requiring runtime reconfiguration to change S. In the actual implementation, 2 PEs were implemented in 6 slices since sharing of resources between adjacent PEs was necessary in the actual implementation.

Fig. 23.3 shows the mapping of the PEs to slices. All the signals ending with "_1st" were used in PE Number 1, and signals ending with "_2nd" were used for PE2. The purpose of each signal can be understood by referring back to Fig. 23.2. It was necessary to connect the output of the RAM-based flip-flops directly to a flip-flop (FF in the diagram) in the same logic cell (LC) since internal LC connections do not permit them to be used independently (i.e. it was not possible to avoid connecting the output of the RAM and the input of the FF). Thus, Slice 1 was configured as a 2 stage shift register for consecutive values of S_i and Slice 3 was used for two consecutive values of T_i.

Fig. 23.4 shows the overall system data path. Since the T sequences are shifted continuously, the system used a FIFO constructed from Block RAM to buffer the sequence data supplied by a host computer. This improves throughput

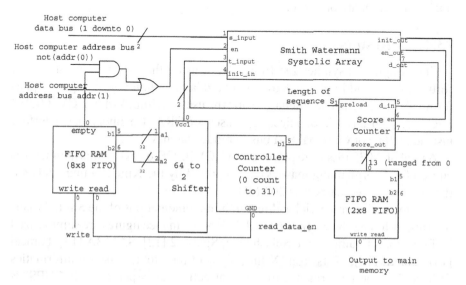

Figure 23.4. System data path.

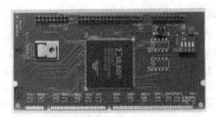

Figure 23.5. Photograph of the Pilchard board.

of the system since a large number of string comparisons can be completed before all of their scores are read from the controller, reducing the amount of idle time in the systolic array. The input and output data width of the FIFO RAM were both 64 bits. The wide input data width helped to improve IO bandwidth from the host computer to the FIFO RAM. A 64-to-2 shifter and a controller counter were used for reducing the output data width of the FIFO RAM from 64 bits to 2 bits, so as to allow data to be fed into the systolic array.

The Score Counter computes the edit distance by accumulating results calculated in the last PE of the systolic array. The output of the last PE is actually the d value in the squares of the rightmost column of the matrix, and differences in values of consecutive squares in the rightmost column must be 1. The $data_{out}$ of the last PE is '0' when $d = b - 1$, and the output '1' when $d = b + 1$. Therefore, a Shift Counter was initialized to the length of the sequence S. It is decremented if the output value was '0', otherwise it is incremented. After the entire string T is passed through the systolic array, the counter contains the final string comparison score.

23.4   Results

The design was synthesized from VHDL using the Xilinx Foundation 5.1i software tools and implemented on Pilchard, a reconfigurable computing platform [8] (Fig. 23.5). The Pilchard platform uses a Xilinx Virtex XCV1000E-6 FPGA (which has 12,288 slices) and uses a SDRAM memory bus interface instead of the conventional PCI bus to reduce latency.

A total of 4,032 PEs were places on an XCV1000E-6 device (this number was chosen for floorplanning reasons). As reported by the Xilinx timing analyzer, the maximum frequency was 202 MHz.

A number of commercial and research implementations of the Smith Waterman algorithm has been reported and their performance figures are summarized in Table 23.2. Examples are Splash [11], Splash 2 [12], SAMBA [13], Paracel [14], Celera [15], JBits from Xilinx [6], and the HokieGene Bioinformatics Project [7]. The performance measure of cell updates per second (CUPS) is widely used in the literature and hence adopted for our results.

Table 23.2. Performance and hardware size comparison of previous implementations (processor core not including system overheads). Device performance is measured in cell updates per second (CUPS).

System	Number of chips	PEs per chip	System performance (CUPS)	Device performance (CUPS)	Run-time reconfiguration requirement
Splash(XC3090)	32	8	370 M	11 M	No
Splash 2(XC4010)	16	14	43 B	2,687 M	No
SAMBA(XC3090)	32	4	1,280 M	80 M	No
Paracel(ASIC)	144	192	276 B	1,900 M	N/A
Celera (software implementation)	800	1	250 B	312 M	N/A
JBits (XCV1000-6)	1	4,000	757 B	757 B	Yes
JBits (XC2V6000-5)	1	11,000	3,225 B	3,225 B	Yes
HokieGene (XC2V6000-4)	1	7000	1,260 B	1,260 B	Yes
This implementation (XCV1000-6)	1	4,032	742 B	742 B	No
This implementation (XCV1000E-6)	1	4,032	814 B	814 B	No

Splash contains 746 PEs in a Xilinx XC3090 FPGA performing the Smith-Waterman Algorithm. Splash 2's hardware was different from Splash, which used XC4010 FPGAs with a total of 248 PEs. SAMBA [13] incorporated 16 Xilinx XC3090 FPGAs with 128 PEs altogether dedicated to the comparison of biological sequences.

ASIC and software implementations have also been reported. Paracel, Inc. used a custom ASIC approach to do the sequence alignment. Their system used 144 identical custom ASIC devices, each containing approximately 192 processing elements. Celera Genomics Inc. reported a software based system using an 800 node Compaq Alpha cluster.

Both the JBits and the HokieGene Bioinformatics Project were the latest reported sequence alignment systems using the Smith-Waterman Algorithm and use the same PE design. JBits reported performance for two different FPGA chips, the XCV1000-6 and the XC2V6000-5. The HokieGene Bioinformatics Project used an XCV6000-4. As can be seen from the table, the performance of the proposed design is similar to the JBits design on the same size FPGA (a XCV1000-6), and the JBits and HokieGene implementations on an XCV6000 gain performance by fitting more PEs on a chip, and our performance on the same chip would be similar.

300

The implementation was successfully verified using the Pilchard platform whcih provides a 133 MHz, 64-bit wide memory mapped bus to the FPGA. The processing elements and all other logic of the implementation operate from the same 133 MHz clock. The interface logic occupied 3% of the Virtex device. The working design was used mainly for verification purposes and had a disappointing performance of approximately 136 B CUPS, limited by the simple polling based host interface used. A high speed interface which performs more buffering and is able to cause the memory system to perform block transfers between the host and Pilchard is under development.

23.5   Conclusion

A technique, commonly used in VLSI layout, in which two processing elements are merged into a compact cell was used to develop a Smith-Waterman systolic processing element design which computes the edit distance between two strings. This cell occupies 3 Xilinx Virtex slices and allows both strings to be loaded into the system without runtime reconfiguration. Using this cell, 4032 PEs can fit on a Xilinx XCV1000E-6, operate at 202 MHz and achieve a device performance of 814 B CUPS.

References

[1] Y. Miki, et. al. A Strong Candidate for the Breast and Ovarian Cancer Susceptibility Gene, BRCA1. *Science*, 266:66–71, 1994.
[2] European Bioinformatics Institute Home Page, FASTA searching program, 2003. http://www.ebi.ac.uk/fasta33/.
[3] National Center for Biotechnology Information. NCBI BLAST home page, 2003. http://www.ncbi.nlm.nih.gov/blast.
[4] T. F. Smith and M. S. Watermann. Identification of common molecular subsequence. *Journal of Molecular Biology*, 147:196–197, 1981.
[5] Sencel's search software, 2003. http://www.sencel.com.
[6] Steven A. Guccione and Eric Keller. Gene matching using JBits, 2002. http://www.ccm.ece.vt. edu/hokiegene/papers/GeneMatching.pdf.
[7] K. Puttegowda, W. Worek, N. Pappas, A. Danapani and P. Athanas. A run-time reconfigurable system for gene-sequence searching. In *Proceedings of the International VLSI Design Conference*, page (to appear), Jan 2003.
[8] P. Leong, M. Leong, O. Cheung, T. Tung, C. Kwok, M. Wong and K. H. Lee. Pilchard—a reconfigurable computing platform with memory slot interface. In *Proceedings of the IEEE Symposium on Field-Programmable Custom Computing Machines*, page (to appear), April 2001.
[9] Richard J. Lipton and Daniel Lopresti. A systolic array for rapid string comparison. In *Proceedings of the Chapel Hill Conference on VLSI*, pages 363–376, 1985.
[10] Xilinx. *The programmable logic data book*, 2003.
[11] D. T. Hoang. A systolic array for the sequence alignment problem. *Brown University, Providence, RI, Technical Report*, pages CS–92–22, 1992.
[12] D. T. Hoang. Searching genetic databases on splash 2. In *Proceedings 1993 IEEE Workshop on Field-Programmable Custom Computing Machines*, pages 185–192, 1993.
[13] Dominique Lavenier. SAMBA: Systolic Accelerators for Molecular Biological Applications, March 1996.
[14] Paracel, inc, 2003. http://www.paracel.com.
[15] Celera genomics, inc, 2003. http://www.celera.com.

Chapter 24

The Effects of Polynomial Degrees
On The Hierarchical Segmentation Method

Dong-U Lee and Wayne Luk[1], John D. Villasenor[2],
Peter Y.K. Cheung[3],

[1] *Department of Computing*
Imperial College, London, United Kingdom
{dong.lee, w.luk}@ic.ac.uk

[2] *Electrical Engineering Department*
University of California, Los Angeles, USA
villa@icsl.ucla.edu

[3] *Department of EEE Imperial College, London, United Kingdom*
p.cheung@ic.ac.uk

Abstract   This chapter presents the effects of polynomial degrees on the hierarchical segmentation method (HSM) for approximating functions. HSM uses a novel hierarchy of uniform segments and segments with size varying by powers of two. This
scheme enables us to approximate non-linear regions of a function particularly
well. The degrees of the polynomials play an important role when approximating
functions with HSM: the higher the degree, the fewer segments are needed to
meet the same error requirement. However, higher degree polynomials require
more multipliers and adders, leading to higher circuit complexity and more delay.
Hence, there is a tradeoff between table size, circuit complexity and delay. We
explore these tradeoffs with four functions: $\sqrt{-\log(x)}$, $x \log(x)$, a high order
rational function and $\cos(\pi x/2)$. We present results for polynomials up to the
fifth order for various operand sizes between 8 and 24 bits.

Introduction

In this chapter, we explore the effects of polynomial degrees when approximating functions with the hierarchical segmentation method (HSM) presented

P. Lysaght and W. Rosenstiel (eds.),
New Algorithms, Architectures and Applications for Reconfigurable Computing, 301–313.
© 2005 *Springer. Printed in the Netherlands.*

in [8]. HSM is based on piecewise polynomial approximations, and uses a novel hierarchy of uniform segments and segments with size varying by powers of two. This scheme enables us to use segmentations that take the non-linear regions of a function into account, resulting in significantly fewer segments than the traditional uniform segmentation.

The degrees of the polynomials play an important role when approximating functions with HSM: the higher the degree, the fewer segments are needed to meet the same error requirement. However, higher degree polynomials require more multipliers and adders, leading to higher circuit complexity and more delay. Hence, there is a tradeoff between table size, circuit complexity and delay. We explore these tradeoffs with the following four non-linear compound and elementary functions:

$$f_1 = \sqrt{-\log(x)} \tag{24.1}$$

$$f_2 = x \log(x) \tag{24.2}$$

$$f_3 = \frac{0.0004x + 0.0002}{x^4 - 1.96x^3 + 1.348x^2 - 0.378x + 0.0373} \tag{24.3}$$

$$f_4 = \cos(\pi x/2) \tag{24.4}$$

where the input x is an n-bit number over $[0, 1)$ of the form $0.x_{n-1}..x_0$. Note that the functions f_1 and f_2 cannot be computed for $x = 0$, therefore we approximate these functions over $(0, 1)$ and generate an exception when $x = 0$. In this work, we implement an n-bit in, n-bit out system. However, the position of the decimal (or binary) point in the input and output formats can be different in order to maximize the precision that can be described. We present results for polynomials up to fifth order for various operand sizes between 8 and 24 bits. We have opted for faithful rounding [3], meaning that the results are rounded to the nearest or next nearest, i.e. they are accurate within 1 ulp (unit in the last place).

The principal contribution of this chapter is a review of HSM and a quantitative analysis of the effects of the polynomial degrees. The novelties of our work include:

- a scheme for piecewise polynomial approximations with a hierarchy of segments;
- quantitative analysis of the effects of using different order polynomials;
- evaluation with four compound functions;
- hardware implementation of the proposed method.

The rest of this chapter is organized as follows: Section 24.1 covers background material and previous work. Section 24.2 reviews HSM. Section 24.3 explores how the polynomial degrees affect HSM at various operand sizes. Section 24.4 discusses evaluation and results, and Section 24.5 offers conclusions and future work.

24.1   Background

Many applications including digital signal processing, computer graphics and scientific computing require the evaluation of mathematical functions. Applications that do not require high precision, often employ direct table look-up methods. However, this can be impractical for precisions higher than a few bits, because the size of the table is exponential in the input size.

Recently, table look-up and addition methods have attracted significant attention. Table look-up and addition methods use two or more parallel table look-ups followed by multi-operand addition. Such methods include the symmetric bipartite table method (SBTM) [14] and the symmetric table addition method (STAM) [15]. These methods exploit the symmetry of the Taylor approximations and leading zeros in the table coefficients to reduce the look-up table size. Although these methods yield significant improvements in table size over direct look-up techniques, they can be inefficient for functions that are highly non-linear.

Piecewise polynomial approximation [11] which is the method we use in this chapter, involves approximating a continuous function f with one or more polynomials p of degree d on a closed interval $[a, b]$. The polynomials are of the form

$$p(x) = c_d x^d + c_{d-1} x^{d-1} + \cdots + c_1 x + c_0 \qquad (24.5)$$

and with Horner's rule, this becomes

$$p(x) = ((c_d x + c_{d-1})x + \cdots)x + c_0 \qquad (24.6)$$

where x is the input. The aim is to minimize the distance $\| p - f \|$. Our work is based on minimax polynomial approximations, which involve minimizing the maximum absolute error. The distance for minimax approximations is:

$$\| p - f \|_\infty = \max_{a \le x \le b} |f(x) - p(x)|. \qquad (24.7)$$

Piñeiro et al. [13] divide the input interval into several uniform segments. For each segment, they store the second degree minimax polynomial approximation coefficients, and accumulate the partial terms in a fused accumulation tree. Such approximations using uniform segments [12], [1] are suitable for functions with linear regions, but are inefficient for non-linear functions, especially when the function varies exponentially. It is desirable to choose the boundaries of the segments to cater for the non-linearities of the function. Highly non-linear regions may need smaller segments than linear regions. This approach minimizes the amount of storage required to approximate the function, leading to more compact and efficient designs. HSM uses a hierarchy of uniform segments (US) and powers of two segments (P2S), that is segments with the size varying by increasing or decreasing powers of two.

Similar approaches to HSM have been proposed for the approximation of the non-linear functions in logarithmic number systems. Henkel [4] divides the interval into four arbitrary placed segments based on the non-linearity of the function. The address for a given input is approximated by another function that approximates the segment number for an input. This method only works if the number of segments is small and the desired accuracy is low. Also, the function for approximating the segment addresses is non-linear, so in effect the problem has been moved into a different domain. Coleman et al. [2] divide the input interval into seven P2S that decrease by powers of two, and employ constant numbers of US nested inside each P2S, which we call P2S(US). Lewis [9] divides the interval into US that vary by multiples of three, and each US has variable numbers of uniform segments nested inside, which we call US(US). However, in both cases the choice of inner and outer segment numbers is done manually, and a more efficient segmentation could be achieved using our segmentation scheme.

24.2 The Hierarchical Segmentation Method

Let f be a continuous function on $[a, b]$, and let an integer $m \geq 2$ specify the number of contiguous segments into which $[a, b]$ has been partitioned: $a = u_0 \leq u_1 \leq \ldots \leq u_k = b$. Let d be a non-negative integer and let P_i denote the set of functions p_i whose polynomials are of degree less or equal to d. For $i = 1, \ldots, m$, define

$$h_i(u_{i-1}, u_i) = \min_{p_i \in P_i} \max_{u_{i-1} \leq x \leq u_i} |f(x) - p_i(x)|. \qquad (24.8)$$

Let $e_{max} = e_{max}(u) = \max_{1 \leq i \leq m} h_i(u_{i-1}, u_i)$. The segmented minimax approximation problem is that of minimizing e_{max} over all partitions u of $[a, b]$. If the error norm is a non-decreasing function of the length of the interval of approximation, the function to be approximated is continuous and the goal is to minimize the maximum error norm on each interval, then a balanced error solution is optimal. The term "balanced error" means that the error norms on each interval are equal [5].

We presented an algorithm to find these optimum boundaries in [8], based on binary search. This algorithm is implemented in MATLAB, where the function to be approximated f, the interval of approximation $[a, b]$, the degree d of the polynomial approximations and the number of segments m are given as inputs. The program outputs the segment boundaries $u_{1..m-1}$ and the maximum absolute error e_{max}. Figure 24.1 shows the optimum segment boundaries for 16 and 24-bit second order approximations to f_1. We observe that f_1 needs more segments in the regions near zero and one, i.e. smaller segments are needed when the curve has rapidly changing non-linear regions.

In the ideal case, one would use these optimum boundaries to approximate the functions. However, from a hardware implementation point of view, this can

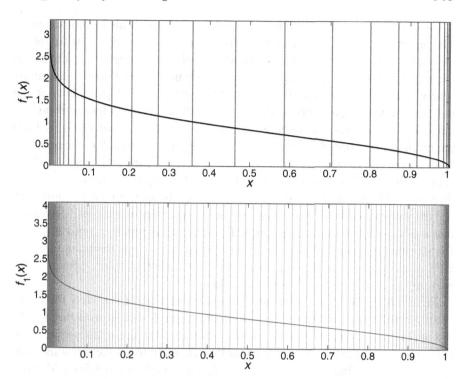

Figure 24.1. Optimum locations of the segments for 16 and 24-bit operands and second order approximations to f_1, resulting in 4, operands and second order approximations to f_1, resulting in 44 and 358 segments respectively.

be impractical. The circuit to find the right segment for a given input could be complex, hence large and slow. A more hardware-friendly systematic segmentation scheme is needed which leads to the development of HSM. Nevertheless, the optimum segments give us an indication of how well a given segmentation scheme matches the optimum segmentation in terms of numbers of segments. Moreover, they provide information on the non-linearities of a function.

HSM, which is based on piecewise polynomial approximations and uses an efficient hierarchy of US and P2S is also presented in [8]. This scheme enables us to use segmentations that take the non-linear regions of a function into account, resulting in significantly fewer segments than the traditional uniform segmentation. The hierarchy schemes H we have chosen are P2S(US), P2SL(US), P2SR(US) and US(US). These four schemes cover most of the non-linear functions of interest. P2S means segments that increase by powers of two up to the mid-point then decrease by powers of two. P2SL are segments that increase by powers of two, and P2SR are segments that decrease by powers of two. US are the conventional uniform segments. P2SL(US) would mean that the outer segmentation is P2SL and there are inner segments US nested inside the outer segments.

We have implemented HSM in MATLAB, which deals with the four schemes. The program called HFS (hierarchical function segmenter) takes the following inputs: the function f to be approximated, the polynomial degree d, input range, the requested output error e_{max}, ulp of the input, operand size n, hierarchy scheme H, number of bits for the outer segment v_0, and precisions of the polynomial coefficients and the data paths. HFS divides the input interval into outer segments whose boundaries are determined by H and v_0. HFS finds the optimum number of bits v_1 for the inner segments for each outer segment, which meets the requested output error constraint. For each outer segment, HFS starts with $v_1 = 0$ and computes the error e of the approximation. If $e > e_{max}$ then v_1 is incremented and the error e for each inner segment is computed, i.e. the number of inner segments is doubled in every iteration. If it detects that $e > e_{max}$ it increments v_1 again. This process is repeated until $e \leq e_{max}$ for all inner segments of the current outer segment. This is the point at which HFS obtains the optimum number of bits for the current outer segment. HFS performs this process for all outer segments.

Figure 24.2 shows the segmented functions obtained from HFS for 16-bit second order approximations to the four functions. It can be seen that for f_3, the segments produced by HFS closely resemble the optimum segments in Figure 24.1. Double precision is used for the data paths to get these results. Note that for relatively linear functions such as f_4, the advantages of using HSM over uniform segmentation is small. The advantages of HSM become more apparent as the functions become more non-linear (f_1 being an extreme example).

24.3 The Effects of Polynomial Degrees

The degrees of the polynomials play an important role when approximating functions with HSM: the higher the degree, the fewer segments are needed to meet the same error requirement. However, higher degree polynomials require more multipliers and adders, leading to higher circuit complexity and more delay. Hence, there is a tradeoff between table size, circuit complexity and delay.

Table size. As the polynomial degree d increases, the width of the coefficient look-up table increases linearly. One needs to store $d + 1$ coefficients per segment.

Circuit Complexity. As d increases, one needs more adders and multipliers to perform the actual polynomial evaluation. These increase in a linear manner: since we are using Horner's rule, d adders and d multipliers are required.

Delay. Note that the polynomial coefficients in the look-up table can be accessed in parallel. Hence the delay differences between the different polynomial degrees occur when performing the actual polynomial evaluation. Due to the serial nature of Horner's rule, the increase in delay of the polynomial evaluation is again linear (d adders and d multipliers).

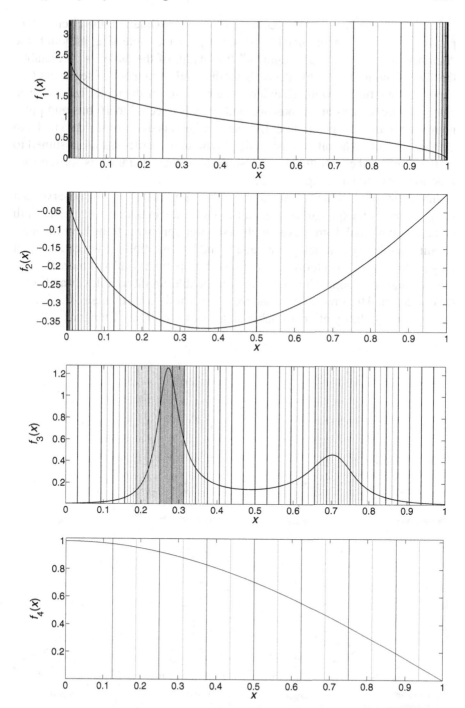

Figure 24.2.   The segmented functions generated by HFS for 16-bit second order approximations. f_1, f_2, f_3 and f_4 employ P2S(US), P2SL(US), US(US) and US(US) respectively. The
black and grey vertical lines are the boundaries for the outer and inner segments respectively.

We vary the polynomial degree for a given approximation, the circuit complexity and delay are predictable. However, for the table size, although the "number of coefficients per segment" (the width of the table) is predictable, the size of the look-up table depends on the total of number of segments (the depth of the table) as well. To explore the behavior of the table size, we have calculated table sizes at various operand sizes between 8 to 16 bits and polynomial degrees from one to five. Double precision is used for the data paths to get these results. The bit widths of the polynomial coefficients are assumed to be the same as the operand size. Mesh plots of these parameters for the four functions are shown in Figure 24.3.

Interestingly, all four functions share the same behavior. We observe that the plots have an exponential behavior in table size in both the operand width and the polynomial degree. Although first order approximations have the least circuit complexity and delay (one adder and one multiplier), we can see that they perform poorly (in terms of table size) for operand widths of more than 16 bits. Second order approximations have reasonable circuit complexity and delay (two adders and two multipliers) and for the bit widths used in these experiments (up to 24 bits), they yield reasonable table sizes for all four functions.

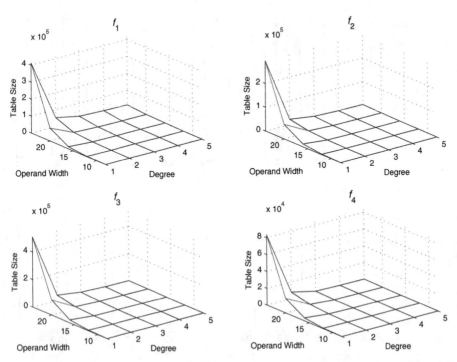

Figure 24.3. Variations of the table sizes to the four functions with varying polynomial degrees and operand bit widths.

Another observation is that the improvements in table size of using third or higher order polynomials are very small. For instance, looking at the 24-bit results to f_2, the table sizes of first, second and third order approximations are 294768, 22680 and 8256 bits respectively. The difference between first and second order is a factor of 11.3, whereas the difference between second and third order is a factor of 2.7. Therefore, the overhead of having an extra adder and multiplier stage for third order approximations maybe not be worth while for a table size reduction of a factor of just 2.7.

Hence, we conclude that for operand sizes of 16 bits or fewer, first order approximations yield good results. For operand sizes between 16 and 24 bits, second order approximations are perhaps more appropriate. We predict that for operand sizes larger than 24 bits, the table size improvements of using third or higher order polynomials will appear more dramatic.

In [8], we have compared the number of segments of HSM to the optimum segmentation. For first and second order approximations, the ratio of segments obtained by HSM and the optimum is around a factor of two. To explore how this ratio behaves at varying operand widths and polynomial degrees, we obtain the results shown in Figure 24.4. We can see that the ratios are around a factor of two at various parameters, with HSM showing no obvious signs of degradation.

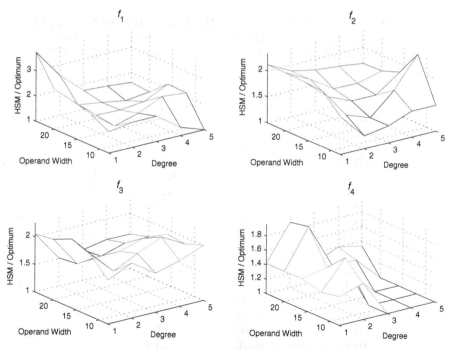

Figure 24.4. Variations of the HSM / Optimum segment ratio with polynomial degrees and operand bit widths.

24.4 Evaluation and Results

Table 24.1 compares HSM with direct table look-up, SBTM and STAM for 16 and 24-bit approximations to f_2. We observe that table sizes for direct look-up approach are not feasible when the accuracy requirement is high. SBTM/STAM significantly reduce the table sizes compared to the direct table look-up approach, at the expense of some adders and control circuitry. In the 16-bit results, HSM_4 has the smallest table size, being 546 and 8.5 times smaller than direct look-up and STAM. The table size improvement of HSM is of course at the expense of more multipliers and adders, hence higher latency. Generally, the higher the polynomial degree, the smaller the table size. However, in the 16-bit case, HSM_5 actually has larger table size than HSM_4. This is because of the extra overhead of having to store one more polynomial coefficient per segment exceeds the reduction in number of segments compared to HSM_4. We observe for the 24-bit results, the differences in table sizes between HSM and other methods are even larger. Moreover, the reductions in table size by using higher order polynomials get greater as the operand widths increase (i.e. as the accuracy requirement increases). For applications that require relatively low accuracies and latencies, SBTM/STAM may be preferred. For high accuracy applications that can tolerate longer latencies, HSM would be more appropriate.

Table 24.1. Comparison of direct look-up, SBTM, STAM and HSM for 16 and 24-bit approximations to f_2. The subscript for HSM denotes the polynomial degree, and the subscript for STAM denotes the number of multipartite tables used. Note that SBTM is equivalent to $STAM_2$.

operand width	method	table size (bits)	compression	multiplier	adder
16	direct	1,048,576	546.1	—	—
	SBTM	29,696	15.5	—	1
	$STAM_4$	16,384	8.5	—	3
	HSM_1	24,384	12.7	1	2
	HSM_2	4,620	2.4	2	3
	HSM_3	2,304	1.2	3	4
	HSM_4	1,920	1.0	4	5
	HSM_5	2,112	1.1	5	6
24	direct	402,653,184	77,672.3	—	—
	SBTM	2,293,760	442.5	—	1
	$STAM_6$	491,520	94.8	—	5
	HSM_1	393,024	75.8	1	2
	HSM_2	40,446	7.8	2	3
	HSM_3	11,008	2.1	3	4
	HSM_4	6,720	1.3	4	5
	HSM_5	5,184	1.0	5	6

Table 24.2. Hardware synthesis results on a Xilinx Virtex-II XC2V4000-6 FPGA for 16 and 24-bit, first and second order approximations to f_2 and f_3.

function	order	operand width	speed (MHz)	latency (cycles)	slices	block RAMs	block multipliers
f_2	1	16	202	11	332	2	2
		24	160	12	897	44	4
	2	16	153	13	483	1	4
		24	135	14	871	2	10
f_3	1	16	232	8	198	2	1
		24	161	10	418	37	2
	2	16	198	12	234	1	3
		24	157	13	409	3	4

The reference design templates have been implemented using Xilinx System Generator. These design templates are fully parameterizable, and changes to the desired function, input interval, operand width or finite precision parameters can result in producing a new design automatically.

A variant [7] of our approximation scheme to f_1 and f_4, with one level of P2S and US(P2S), has been implemented and successfully used for the generation of Gaussian noise samples [6]. Table 24.2 contains implementation results for 16 and 24-bit, first and second order approximations to f_2 and f_3, which are mapped and tested on a Xilinx Virtex-II XC2V4000-6 FPGA. The bit widths of the coefficients and the data paths have been optimized by hand to minimize the size of the multipliers and look-up tables. The design is fully pipelined generating a result every clock cycle. Designs with lower latency and clock speed can be obtained by reducing the number of pipeline stages. The designs have been tested exhaustively over all possible input values to verify that all outputs are indeed faithfully rounded.

We can see that for the 16-bit results, the first order approximations have faster speed, lower latency, and fewer slices and block multipliers. But they use slightly more block RAMs, which is due to the larger numbers of segments required. For the 24-bit cases, the first order results are again faster (in terms of both clock speed and latency) and use fewer block multipliers, but they use slightly more slices and many more block RAMs. This is due the inefficiency of the first order approximations for large operand bit widths as discussed in Section 24.3.

Our hardware implementations have been compared with software implementations (Table 24.3). The FPGA implementations compute the functions using HSM with 24-bit operands and second order polynomials. Software implementations are written in C generating single precision floating point numbers, and are compiled with the GCC 3.2.2 compiler. This is a fair comparison in terms of precision, since single precision floating point has 24-bit mantissa accuracy.

Table 24.3. Performance comparison: computation of f_2 and f_3 functions. The XC2V4000-6 FPGA belongs to the Xilinx Virtex-II family. The Athlon and the Pentium PCs are equipped with 512MB and 1GB DDR RAMs respectively.

function	platform	speed (MHz)	throughput (operations/second)	completion time (ns)
f_2	XC2V4000-6 FPGA	135	135 million	104
	AMD Athlon PC	1400	7.14 million	140
	Intel Pentium 4 PC	2600	0.48 million	2088
f_3	XC2V4000-6 FPGA	157	157 million	83
	AMD Athlon PC	1400	1.76 million	569
	Intel Pentium 4 PC	2600	1.43 million	692

For the f_2 function, the Virtex-based FPGA implementation is 20 times faster than the Athlon-based PC in terms of throughput, and 1.3 times faster in terms of completion time. We suspect that the inferior results of the Pentium 4 PC are due to inefficient implementation of the log function in the gcc math libraries for the Pentium 4 CPU. Looking at the f_3 function, the FPGA implementation is 90 times faster than the Athlon-based PC in terms of throughput, and 7 times faster in terms of completion time. This increase in performance gap is due to the f_3 function being more 'compound' than the f_2 function. Whereas a CPU computes each elementary operation of the function one by one, HSM looks at the entire function at once. Hence, the more compound a function is, the advantages of HSM get bigger.

Note that the FPGA implementations use only a fraction (less than 2%) of the device used, hence by instantiating multiple function evaluators on the same chip for parallel execution, we can expect even larger performance improvements.

24.5 Conclusion

In this chapter, we have presented the effects of polynomial degrees on the hierarchial segmentation method (HSM) for evaluating functions. The effects are explored using four functions: $\sqrt{-\log(x)}$, $x \log(x)$, a high order rational function and $\cos(\pi x/2)$. HSM uses a novel hierarchy of uniform segments and segments with size varying by powers of two. This scheme enables us to approximate non-linear regions of a function particularly well. Compared to other popular methods such as STAM, our approach has longer latencies and operators, but the size of the look-up tables are considerably smaller. The table size improvements get larger as the operand bit width increases.

Looking at the effects of the polynomial degrees, we have seen that first order approximations are appropriate for operand widths of fewer than 16 bits, and second order approximations are well suited for operand widths between

16 and 24 bits. We expect that the advantages of using higher order polynomials will become larger for operands widths of more than 24 bits. We have also seen that the performance of HSM over the optimum segmentation is around a factor of two throughout various operand widths and polynomial degrees.

References

[1] J. Cao, B.W.Y. We and J. Cheng, "High-performance architectures for elementary function generation", *Proc. 15th IEEE Symp. on Comput. Arith.*, 2001.

[2] J.N. Coleman, E. Chester, C.I. Softley and J. Kadlec, "Arithmetic on the European logarithmic microprocessor", *IEEE Trans. Comput. Special Edition on Comput. Arith.*, vol. 49, no. 7, pp. 702–715, 2000.

[3] D. Das Sarma and D.W. Matula, "Faithful bipartite rom reciprocal tables", *Proc. IEEE Symp. on Computer Arithmetic*, pp. 17–28, 1995.

[4] H. Henkel, "Improved Addition for the Logarithmic Number System", *IEEE Trans. on Acoustics, Speech, and Signal Process.*, vol. 37, no. 2, pp. 301–303, 1989.

[5] C.L. Lawson, "Characteristic properties of the segmented rational minimax approximation problem", *Numer. Math.*, vol. 6, pp. 293–301, 1964.

[6] D. Lee, W. Luk, J. Villasenor and P.Y.K. Cheung, "A hardware Gaussian noise generator for channel code evaluation", *Proc. IEEE Symp. on Field-Prog. Cust. Comput. Mach.*, pp. 69–78, 2003.

[7] D. Lee, W. Luk, J. Villasenor and P.Y.K. Cheung, "Hardware function evaluation using non-linear segments", *Proc. Field-Prog. Logic and Applications*, LNCS 2778, Springer-Verlag, pp. 796–807, 2003.

[8] D. Lee, W. Luk, J. Villasenor and P.Y.K. Cheung, "Hierarchical Segmentation Schemes for Function Evaluation", *Proc. IEEE Int. Conf. on Field-Prog. Tech.*, pp. 92–99, 2003.

[9] D.M. Lewis, "Interleaved memory function interpolators with application to an accurate LNS arithmetic unit", *IEEE Trans. on Comput.*, vol. 43, no. 8, pp. 974–982, 1994.

[10] O. Mencer, N. Boullis, W. Luk and H. Styles, "Parameterized function evaluation for FPGAs", *Proc. Field-Prog. Logic and Applications*, LNCS 2147, Springer-Verlag, pp. 544–554, 2001.

[11] J.M. Muller, *Elementary Functions: Algorithms and Implementation*, Birkhauser Verlag AG, 1997.

[12] A.S. Noetzel, "An interpolating memory unit for function evaluation: analysis and design", *IEEE Trans. Comput.*, vol. 38, pp. 377–384, 1989.

[13] J.A Piñeiro, J.D. Bruguera and J.M. Muller, "Faithful powering computation using table look-up and a fused accumulation tree", *Proc. 15th IEEE Symp. on Comput. Arith.*, 2001.

[14] M.J. Schulte and J.E. Stine, "Symmetric bipartite tables for accurate function approximation", *Proc. 13th IEEE Symp. on Comput. Arith.*, vol. 48, no. 9, pp. 175–183, 1997.

[15] J.E. Stine and M.J. Schulte, "The symmetric table addition method for accurate function approximation", *J. of VLSI Signal Processing*, vol. 21, no. 2, pp. 167–177, 1999.